# Advances in Dielectric Elastomer Composites

This is a comprehensive guide to dielectric elastomer composites (DECs), which play an integral role in new electromechanically active smart-material technologies.

Introducing the basic concepts behind DECs, the book is accessible to readers at all levels. It demonstrates how to implement practical problem-solving methods through nonlinear continuum mechanics and also discusses fiber-reinforced anisotropic DE composites and the electromechanically coupled behavior of anisotropic fiber-reinforced DEs. Using practical examples throughout, it proposes models which utilize the strain energy function, nonlinear electro-elasticity, and nonlinear continuum mechanics. It presents clear guidelines for creating practical nonlinear finite element code, and introduces the reader to hyperelasticity.

The book will be an accessible reference guide for students, researchers, and engineers in the field of mechanical engineering, bioengineering, materials science, aeronautics, and robotics.

# Advances in Dielectric Elastomer Composites

## A Nonlinear Elasticity Framework

Masoud Asgari
Marzie Majidi

CRC Press is an imprint of the
Taylor & Francis Group, an **informa** business

Front cover image: Shutterstock

First edition published 2025
by CRC Press
2385 NW Executive Center Drive, Suite 320, Boca Raton FL 33431

and by CRC Press
4 Park Square, Milton Park, Abingdon, Oxon, OX14 4RN

*CRC Press is an imprint of Taylor & Francis Group, LLC*



ISBN: 978-1-032-94564-4 (hbk)
ISBN: 978-1-032-94569-9 (pbk)
ISBN: 978-1-003-57146-9 (ebk)

DOI: 10.1201/9781003571469

Typeset in Times
by SPi Technologies India Pvt Ltd (Straive)

# Contents

# Figures

# Tables

# Preface

The rapidly growing needs and applications for efficient soft and smart actuators and transducers having high performance has resulted in significant efforts for developing devices based on materials with electroactive properties. Electromechanically active polymers (EAP) are a main candidate within the smart materials due to their great characteristic potential as well as the cost and versatility of elastomeric materials. Among the EAPs, the specific field of dielectric elastomers (DE) is of considerable importance regarding its overall performance, simplicity, robustness and fast-growing applications as biomedical implants, sensors, energy harvesters and soft robots. Meanwhile, more possibilities could be explored by incorporating other properties of DEs including viscoelasticity and anisotropy.

This book provides a comprehensive and organized overview to the specific field of dielectric elastomer composites (DEC) as an important category of the emerging electromechanically active smart-material technology. The aim is to describe basic concepts as well as implement practical problem-solving methods based on nonlinear continuum mechanics foundation in an accessible form. The main concern of the book is multi-layered and fiber-reinforced anisotropic DE composites through a step by step description of governing equations. The text focuses on establishing constitutive laws to capture the electromechanically coupled behavior of anisotropic fiber-reinforced DEs. As practical examples, some proposed models are presented in the framework of the strain energy function, nonlinear electro-elasticity, and nonlinear continuum mechanics approach. Their stepwise explanations make this book an easy-to-use reference guide for students, researchers and engineers.

The theoretical foundation beside numerical implementation techniques in separate chapters addressing the modelling and simulations will lead to applicable procedures on how to get started in design of innovative DEC structures and actuators. The functional and structural properties of anisotropic DE composites are described and explained in basic nonlinear finite elasticity approach via various illustrative case studies.

It is not just another book on EAPs. It is intended to be a comprehensive reference for focusing on the advanced types of multi-layered and anisotropic DECs in order to respond a critical gap between the scientific literature and classic theoretical textbooks. It provides a comprehensive and organized text of advanced topics at a tutorial level including basic concepts and governing equations in a simple form, as well as modelling guidelines for practical implantations.

# Author Biography

**Masoud Asgari** Masoud Asgari is an Associate Professor of Mechanical Engineering at K. N. Toosi University of Technology in Tehran, Iran. He has made significant contributions to both academia and industry. Dr. Asgari earned his PhD from Amirkabir University of Technology (Tehran Polytechnic) with distinction, following his excellent undergraduate and master's studies at Sharif University of Technology. Both his dissertation and thesis received the highest honors, reflecting his dedication and academic excellence. His research focuses on Lightweight Construction, Additive Manufacturing, and the development of Functional Materials. Dr. Asgari's contributions have gained international recognition, with research stays at prestigious institutions such as the University of Modena and Reggio Emilia (Italy) and Bern University of Applied Sciences (Switzerland). In recognition of his impactful work, he has been named in the Stanford-Elsevier Ranking of the Top 2% Highly Cited Scientists in 2022 and 2023. In 2024, he was awarded a prestigious Alexander von Humboldt Research Fellowship for experienced researchers. Dr. Asgari also serves as a reviewer for leading journals indexed in the Web of Science, contributing to the advancement of knowledge in Mechanical Engineering.

**Marzie Majidi** received her Ph.D. degree in mechanical engineering K. N. Toosi University of Technology in Tehran, Iran, in 2023. Her research focused on modeling dielectric and magnetorheological elastomers, key materials for innovative actuators. Dr. Majidi is currently at Saipa Automative Industries Research Innovative Center (AIRIC), working on Noise, Vibration, and Harshness (NVH) analysis as a CAE expert. She received the M.Sc. and B.Sc. degrees in mechanical engineering from University of Tehran (UT), in 2017, and Buali Sina university, in 2014. Her research has been on modeling of dielectric and magnetorheological elastomers, Topological Interlocking Material (TIMs), finite element methods, and automative NVH analysis.

# List of Symbols

| | |
|---|---|
| $A$ | Area |
| $A^{(i)}, a^{(i)}$ | Unit Vector Along the i[th] Fibers Family |
| $\mathcal{B}, \mathcal{B}_0$ | Continuum Body in Reference and Current Configurations |
| $B$ | Left Cauchy-Green Deformation Tensor |
| $C$ | Capacity |
| $C$ | Right Cauchy-Green Deformation Tensor |
| $c$ | Damping Coefficient |
| $C_{ij}$ | Constants of Hyperelastic Model |
| $\mathbb{C}, \mathbb{c}$ | Elasticity Tensor in Reference and Current Configurations |
| $D$ | Rate of Deformation |
| $d$ | Thickness |
| $d_{(i)}$ | The *i*th Layer's Thickness |
| $df$ | Force Vector on The Deformed Body |
| $dS, ds$ | Reference and Current Surface Vector |
| $\mathbb{D}, \mathbb{d}$ | Nominal and True Electric Displacement Vectors |
| **DDSDDE** | Consistent Tangent Modulus In Abaqus |
| $\mathcal{D}$ | Dissipation Potential Function |
| $E$ | Green Strain Tensor |
| $\mathbb{E}, \mathbb{e}$ | Nominal and True Electric Field Vectors |
| $F$ | External Force |
| $F$ | Deformation Gradient Tensor |
| $f$ | Body Force |
| $F_z$ | Axial Load |
| $H, h$ | Thickness of Cylinder |
| $h_i(t)$ | Internal Stresses of The Maxwell Elements |
| $I$ | Unity Tensor |
| $I_i, i = 1,2,3$ | Strain Invariants |
| $I_i, i = 4,\ldots,9$ | Pseudo-Invariants Related to Anisotropy |
| $J$ | Jacobian |
| $J_{\text{lim}}$ | Chain Stretch Limit |
| $J_i, i = 1\ldots4$ | Invariants Related to Rate Dependency |
| $\mathrm{K}$ | Kinetic Energy |
| $K$ | Jacobian Matrix |
| $k_1^{(i)}, k_2^{(i)}$ | HGO Model's Constants for the i[th] Fibers Family |
| $l$ | Spatial Velocity Gradient Tensor |
| $l_i, L_i, i = 1,2,3$ | Initial and Final Dimension of An Elastomer |
| $L, l$ | Length of Cylinder |
| $\mathcal{L}$ | Lagrangian |
| $N$ | Fiber Families Number |
| $N'$ | Maxwell Elements Number |
| $N, n$ | Normal Unique Vectors in Reference and Current Configurations |

| | |
|---|---|
| $\tilde{N}$ | Shape Function |
| $n$ | Ogden Model's Rank |
| $\mathbb{N}$ | Number of Layers of a Multi-Layered Structure |
| $P$ | First Piola-Kirchhoff Stress Tensor |
| $P$ | An Arbitrary Particle of The Continuum Body |
| $P_{\text{Maxwell}}$, | Maxwell Stress |
| $p$ | Internal Pressure |
| $\hat{p}$ | Penalty Term |
| $\mathbb{P}, \mathbb{p}$ | Nominal and True Polarization Density |
| $P_i, i = 1,2,3$ | Mechanical Loads |
| $P_r$ | Internal Pressure |
| $Q$ | Electric Charge |
| $R, r$ | Radius of Cylinder |
| $S$ | Second Piola-Kirchhoff Stress Tensor |
| $s_z$ | Actuation Strain |
| $T, t$ | Piola-Kirchhoff (Nominal) Traction Stress and Cauchy (True) Traction Stress |
| $t$ | Time |
| $T_\theta$ | External Torque |
| $\mathcal{T}$ | Nominal Stress |
| $U$ | Potential Energy |
| $u$ | Nodal Displacement Field |
| $u_{\text{elastic}}, u_{\text{electrostatic}}$ | Elastic Strain and Electrostatic Energy Density |
| $v$ | Velocity Field |
| $\mathbb{V}$ | Voltage |
| $V, v$ | Undeformed and Deformed Volume |
| $v_{(m)}$ | Volume Fraction of the m[th] Component |
| $W$ | Helmholtz Free Energy Density |
| $W$ | Spin Tensor |
| $X, x$ | Position Vector In Reference and Current Configurations |
| $Y$ | Young's Modulus of The Elastomer |
| $\alpha$ | Fiber Direction |
| $\alpha_i$ | Dimensionless Model Constants for Ogden Model |
| $\Gamma(t)$ | Stress Relaxation Function |
| $\gamma(t)$ | Normal Stress Relaxation Function |
| $\varepsilon_0, \varepsilon_r$ | Relative Permittivity, Vacuum Permittivity |
| $\epsilon, \epsilon_e, \epsilon_v$ | Total, Elastic and Viscoelastic Deformation of a Rheological Model |
| $\epsilon$ | STRAN in Abaqus |
| $\xi$ | Basic Vector |
| $\eta_i, i = 1,2$ | Rate Dependent Model Constant |
| $\eta_i, i = 1 \ldots N'$ | Viscosity of Dashpot Elements of a Rheological Model |
| $\kappa$ | Bulk Modulus |
| $\hat{\kappa}$ | Temperature Dependent Constant of HGO Model |

| | |
|---|---|
| $\lambda_i, \boldsymbol{i} = \mathbf{1, 2, 3}$ | Principal Stretches |
| $\lambda_a$ | Stretch Along the Unit Vector A |
| $\lambda_f$ | Fiber Stretch |
| $\mu_i$ | Shear Modulus |
| $\mu_j, \boldsymbol{j} = \mathbf{0} \ldots N'$ | Elastic Modulus of Spring Elements of a Rheological Model |
| $\rho$ | Density |
| $\sigma$ | Cauchy Stress Tensor |
| $\tau$ | Kirchhoff Stress Tensor |
| $\tau_0$ | Relaxation Time |
| $\chi$ | Mapping Rule |
| $\Omega$ | Frequency |
| $\Omega_0$ | Reference Configuration |

# Part I

## Fundamentals

# 1 Fundamentals of Dielectric Elastomers

## 1.1 ELECTROACTIVE POLYMERS

Recent technological advances have replaced conventional materials such as metals and alloys with polymers in most industries. Thus, many fields of science and technology have experienced massive revolutions since the polymers' discovery. The successes achieved in polymer technology have also led to the invention of various production methods that provide the possibility of creating desirable properties (mechanical, electrical, etc.) in polymers. These developments have also increased the use of polymers in new fields of science and technology such as the design and production of actuators, sensors, generators, solar cells and memory devices [1].

On the other hand, the rapid development of materials used in industry and technology in recent years has led to the production of materials with intelligence at the molecular level. These smart materials understand the environmental changes and respond to them after processing the information. Polymers that respond to external stimuli by changing their shape or size have been identified and studied over the past decades. These polymers, which react to stimuli such as electric and magnetic fields, chemical agent and light, are known as active polymers. Designed devices using active polymers can mimic the functionality of living systems by generating mechanical, electrical, or optical responses.

Based on the type of stimulation that causes the reaction, whether the change is controllable or reversible, active polymers are classified into different groups. Polymers responsive to electric fields are categorized as Electro Active Polymers (EAPs). Energy transformation, from electrical to mechanical, in addition to flexibility, low density, easy processing, high performance, and low cost [2–6] have led to innovative applications and potentials for EAPs.

The deformation of a polymer material due to the application of an electric field was first observed in 1880 by Roentgen. In his experiment, which was repeated by Sacerdote in 1899, an electric field caused a length change in a rubber strip. Based on his observations, Sacerdote provided an early theory for the strain response to an electric field. Next, in 1925, the first observations in the field of piezoelectric materials were recorded, and in 1969, studies on them began.

Moreover, according to the reaction mechanism, Electroactive Polymers (EAPs) are classified into two classes: ionic EAPs and dielectric EAPs. Figure 1.1 summarizes this classification. Here we go over these classes briefly.

DOI: 10.1201/9781003571469-2

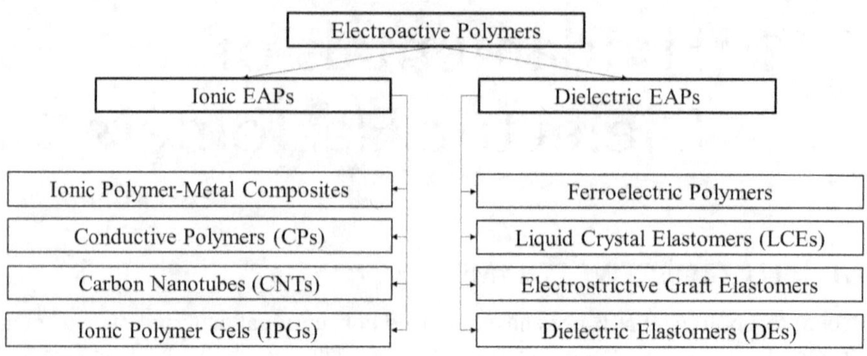


**FIGURE 1.1** Electroactive polymers categories [7].

### 1.1.1 Ionic EAPs

Applying an electric field to the ionic electroactive polymer (IEAPs) will cause a mechanical response through ion transfer or diffusion. This mechanical response is observed as a shape or size change in the material. Created deformation is reversible and the polymer resume its initial configuration as the actuating voltage vanishes. Bending deformation caused by a low electric field is one of the most outstanding types of deformation in these polymers [8].

In addition to low operational voltage, fast response, low power consumption, light weight, flexibility and biological compatibility make IEAPs an efficient alternative for traditional materials and actuators [9].

An IEAP actuator often comprises an electrolyte, as a medium to enable ions immigration, and two flexible plates [8]. Main IEAPs are: Ionic Polymer-Metal Composites (IPMCs), Conductive Polymers (CPs) and Ionic Gels (IGs) [9].

#### 1.1.1.1 Ionic Polymer-Metal Composites (IPMCs)

Ionic Polymer-Metal Composites (IPMCs) actuator is an ion polymer film constrained by two metal electrodes to provide the required electrical field [10]. This actuator is a wet electroactive polymer actuator to provide free movement of cations in the polymer networks and solvent molecules. Consequently, uneven distribution of the solvent leads to IPMC bending toward the anode [4]. The bending direction depends on the polarity. In result of physical move of ions, IPMCs' actuation is slow [10]. Driving mechanisms of ionic polymer-metal composites is presented in Figure 1.2.

Widely used polymers to create IMPCs are perfluorosulfonate (Nafion) and perfluorocarboxylate (Flemion). Common property of these polymers is ion exchange capability which means the polymer can selectively exchange ions of a single charge [11]. Primary materials used for electrodes are gold and platinum because of their high electrochemical stability and electrical conductivity [10].

It is worth noting that liquid polymer composites provide even larger bending actuation and seem promising to enhance the actuator efficiency [10].

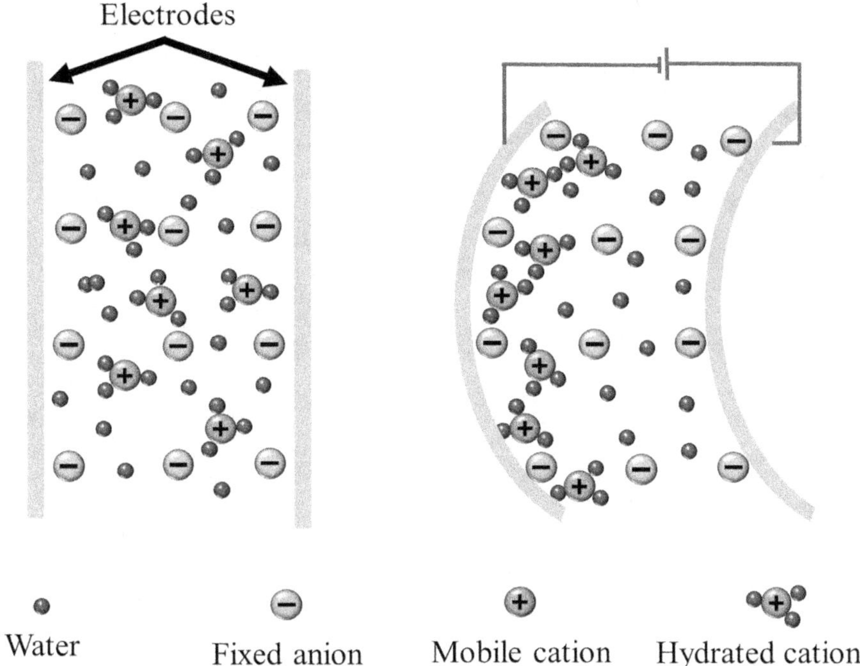


**FIGURE 1.2** Driving mechanism of ionic polymer-metal composites [9].

#### 1.1.1.2 Conductive Polymers (CPs)

Despite the widely accepted fact, some polymers do have the potential to conduct electricity. This property arises from conjugated and alternating single and double carbon-carbon bonds along the polymeric backbone in CPs. CP actuators, consisted of a conductive polymer and an electrolyte, react to electric field by oxidation and reduction. The deformation mechanism is depicted in Figure 1.3. The volume expansion or attraction consequently leads to deformation, mainly bending.

Known polymers with conductivity property include family of polypyrrole (PPy), polyaniline, polythiophene [9]. CP actuators provide large deformation, low cost, and high energy density [12, 13].

Thanks to their valuable properties, a wide range of applications, including various types of sensors [14], drug delivery and bioactuator [15], rechargeable batteries and electrooptical devices have been reported during the recent years.

#### 1.1.1.3 Carbon NanoTubes (CNTs)

In carbon nanotubes, covalently bonded carbons create hollow cylinders. Though several mechanisms have been suggested for their actuation, double-layer charge injection through change of covalent bond seem to be more reasonable. In addition to general properties of electroactive polymers, carbon nanotube benefits from nanoscale structure that suggests them to be used as building blocks in

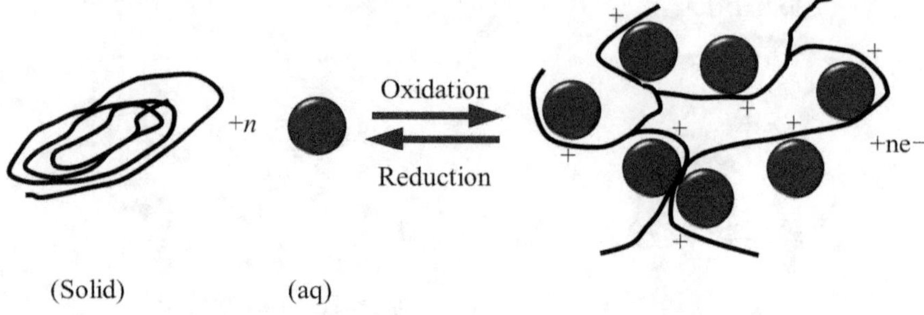


**FIGURE 1.3** Driving mechanism of conductive polymers (CP) [9].

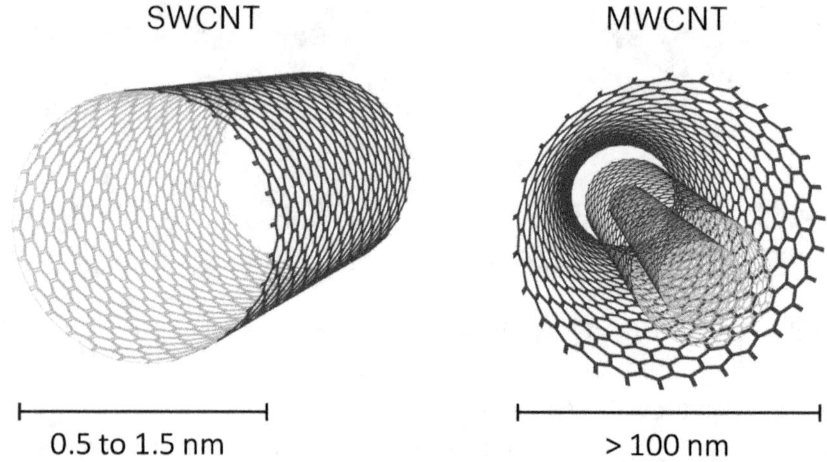


**FIGURE 1.4** Single and multiwall carbon nanotube [18].

nanotechnology [16] and properties that arise from carbon-carbon bonding [17]. Therefore, they are in practice at both the nanoscale (single nanotube) and macroscale (assemblies of nanotubes)[16], each with unique properties and features. As depicted in Figure 1.4, based on the number of walls in the tube, carbon nanotubes are categorized as single-wall carbon nanotubes (SWCNT) or multiwall carbon nanotubes (MWCNT).

Various studies have explored nanotube production techniques, their properties [17], potentials and risks [19], and applications [18].

#### 1.1.1.4 Ion Polymeric Gels (IPGs)

Ion polymeric gel has two phases: a polymeric network and an electrolyte solution. According to the nature of solution, ion gels are aqueous or nonaqueous ionic gels [20]. The gel contains free and fixed ions. In response to chemical or electrical stimuli free ions travel and the gel deforms [21]. Figure 1.5 shows this mechanism.

Common IPGs are polyacrylonitrile, polyacrylic acid (PAA), and poly (methylmethacrylate) [9]. IPG actuators are reported to have potentials to be used in the

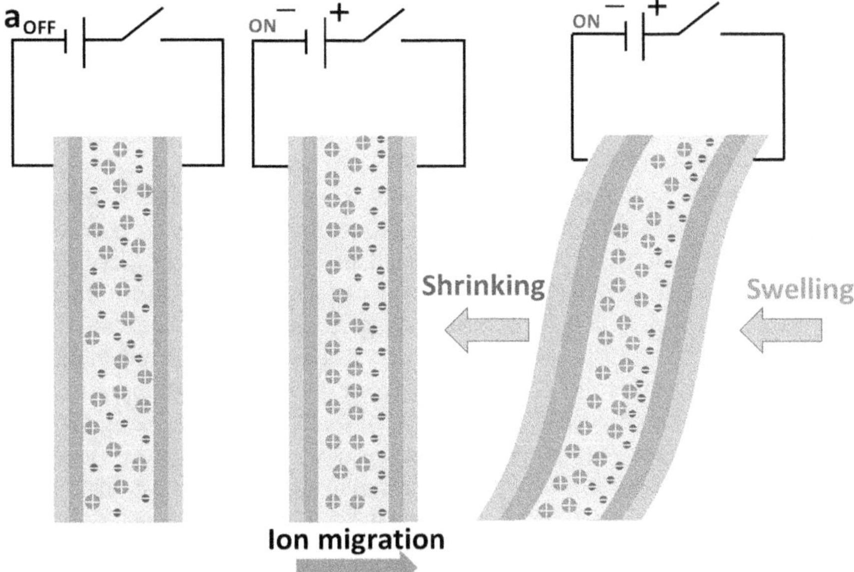


**FIGURE 1.5** Bending mechanism of ion polymeric gels (IPGs) [22].

field of soft robots [23, 24], drug delivery [25] and artificial muscles [22] due to their electromechanical and electrochemical properties.

### 1.1.2 Electronic EAPS

Electronic EAPs actuation mechanism is based on the Coulomb force. Application of an electric field on the two electrodes, sandwiching the polymeric film, move them against each other and transform electrical to mechanical energy. Since there is no need for ion travel to activate the actuator, electronic EAPs work more raped than ionic ones [26].

On the other hand, electronic EAPs need higher voltage to activate because of their low dielectric constant. However, thanks to their ability to hold the induced deformation in their new configuration when activated by DC voltage, they are widely used to devise various robots. Here, we introduce some subgroups of these polymers.

#### 1.1.2.1 Ferroelectric Polymers

Ferroelectric materials exhibit spontaneous electric polarization. The phenomenon is reversible upon applying an electric field [27]. To induce a strain about 2 percent in these polymers, one needs to apply an electric field as large as dielectric breakdown electric field (200 volts per micrometer). Also, due to large young's modulus, ferroelectric polymers have high mechanical energy density. On the other hand, the dielectric losses of these polymers are so significant that challenge their application [28]. Polyvinylidene fluoride (PVDF) known as PVDF is the most popular ferroelectric polymer [29].

The most reported applications of ferroelectric polymers are non-volatile memory devices and data storage [30], energy harvesting, and actuators and sensors [26]. These applications use ferroelectric polymers' properties including easy production, low cost and low weight [1].

#### 1.1.2.2 Liquid Crystal Elastomers (LCEs)

Liquid crystals consist of molecules named mesogens. The properties of these molecules give the tendency for self-organization to the material. They are also polarizable [31]. Since the liquid crystal is an intermediate phase, mesogens may be firmly ordered or fully disordered. Based on the type and degree of mesogens' ordering, three main categories of LCEs have been introduced: nematic, smectic and cholesteric [32]. Nematic LCEs have mesogens aligned across the whole material, while the mesogens of a smectic LCE are arranged in layers and tiled concerning the layer direction. Finally, mesogens of cholesteric LCEs have helical order [26].

Electrostatic energy changes liquid-crystalline side chains and causes stress and mechanical work. The needed electric field to activate LCEs is lower than that in ferroelectric polymers [1]. Energy transformation can happen in the opposite direction as well. Therefore, some actuators and sensors are devised using LCEs [26].

#### 1.1.2.3 Electrostrictive Graft Elastomers

Electrostrictive graft elastomer known as (G-elastomers) and developed by NASA, are a two-constituent system: (1) flexible macromolecular polymer chain (2) grafted crystalline pendant groups [2]. Grafted groups create a 3D elastomer network by forming cross-linking domains. Resulted polarized monomers provide electroactivity in G-elastomers. Application of an electric field make the polar monomers to rotate and align in the field direction. Consequently, the polymer deforms. As the electric field is removed, polar domains attain their random configuration again [1]. Local reorientation of the polymer chain is the second deformation mechanism of G-Elastomers [33]. G-elastomers' structure is as shown in Figure 1.6.

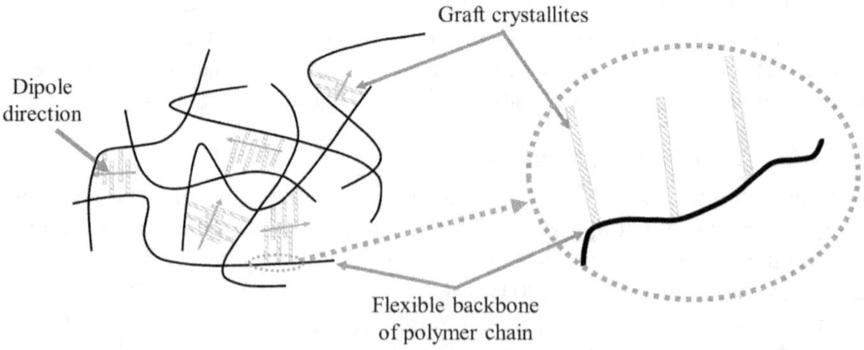


**FIGURE 1.6** Structure of G-elastomers [2].

Since the property of the G-elastomer depends on the ratio of constituents, the desired elongation can be achieved by adjusting this ratio and producing condition [34]. A remarkable property of G-elastomers is their high stiffens which leads to increased output power and design of efficient actuator [33].

## 1.2 HIGH-PERFORMANCE DIELECTRIC ELASTOMERS

Dielectric elastomer is a class of electronic electroactive polymers that exhibits great potential and unique characteristics during the recent decades [7, 35]. DEs are reported to have actuation strains and energy densities of more than 1000% and 3 J/g, respectively. They respond to electric fields in less than one millisecond [36]. Additionally, DEs benefits from inherent compliance, silent operation, low cost [36], high dielectric permittivity and Young's modulus [37].

### 1.2.1 Working Mechanism and Maxwell Stress

In its most straight forward configuration, a planar dielectric elastomer actuator consists of an elastomeric film sandwiched between two compliant thin layers of electrodes [38]. This structure forms a variable capacitor to convert electrostatic energy to mechanical work [36].

Application of a DC electrical voltage upon the two electrodes generates opposite sign charges on them. Therefore, electrodes attract each other because of the electrostatic force between them. As the electrodes move toward each other, the in-between elastomer film compresses in thickness. Figure 1.7 outlines the actuation mechanism of a planar dielectric elastomer actuator.

The stress compressing the elastomer along its direction is known as Maxwell stress, $p$. Actuation strain, $s_z$ (for strains less than 10%) is defined as:

$$s_z = -\frac{P_{\text{Maxwell}}}{Y} \tag{1.1}$$

Y is the Young's modulus of the elastomer. Maxwell stress is approximated as [36]:

$$P_{\text{Maxwell}} = \varepsilon_0 \varepsilon_r \mathbb{E}^2 \tag{1.2}$$

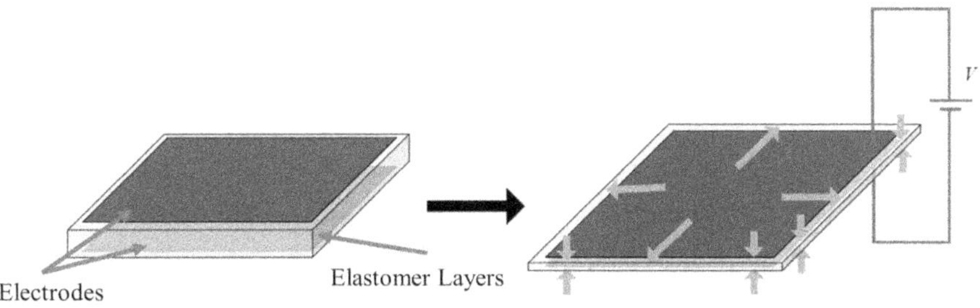


**FIGURE 1.7** Dielectric elastomer actuator working mechanism.

$\varepsilon_r$, $\varepsilon_0$ and $\mathbb{E}$ are relative permittivity, vacuum permittivity ($8.85 \times 10^{-12}$ F/m) and external electric field. The higher dielectric permittivity and lower Young's modulus, the larger actuation strain.

The working mechanism of dielectric elastomer actuator explains conversion of electrical energy to mechanical energy [38]. Thus, a dielectric elastomer is fundamentally a compliant variable capacitor with the capacitance $C$ [39]:

$$C = \frac{\varepsilon_0 \varepsilon_r A}{d} \tag{1.3}$$

$A$ and $d$ are electroded area and film thickness.

Since the elastomer is incompressible (Poisson's ratio = 0.5), its volume remains constant. Therefore, thickness reduction leads to area expansion. This phenomenon was observed and reported by German physics Rontgen in 1880 for the first time [36].

Therefore, the elastic strain energy density of the DEA is $u_{\text{elastic}}$ [36]:

$$u_{\text{elastic}} = -\frac{1}{2} p s_z = \frac{1}{2} Y s_z^2 \tag{1.4}$$

The electrostatic energy density of the DE is $u_{\text{electrostatic}}$ [36]:

$$u_{\text{electrostatic}} = \frac{1}{2} \varepsilon_0 \varepsilon_r E^2 \tag{1.5}$$

The ratio of elastic energy to electrostatic energy density ($\varepsilon_r \varepsilon_0 E^2/Y$) suggests that lower elastic modulus leads to more efficient energy conversion [36]. So, the most important characteristics of the elastomeric film in a DEA are its Young's modulus and permittivity [38].

### 1.2.2 Electrodes

A primary factor influencing the DEs' performance is the quality of electrodes and how they work. Since the elastomeric film undergoes a large deformation, electrodes should follow the elastomer's deformation without causing additional stress. Therefore, they need to be compliant and stretchable [38, 40]. They also have to maintain high conductivity under deformations. The other requirement for electrodes is to be thin and processible to design and fabricate DEAs in various configurations. The electrodes must adhere to the elastomeric film strongly enough to avoid separation during the DEA performance [36, 37].

The most commonly used materials for electrodes are carbon-based electrodes (different deposition forms depicted in Figure 1.8) such as carbon black, graphite, graphene, single-walled carbon nanotubes (SWCNTs), and carbon grease [36] or metallic electrodes such as silver-coated copper flakes [38] and silver nanowires (AgNWs) [36].

**FIGURE 1.8** Different carbon-based electrodes for DEAs [37].

Different electrodes have particular cons and pros. For instance, the oil in the carbon grease gets dried after long exposure times, metallic electrodes have high Young's modulus, or graphene is reported to be a good choice for stacked DEAs [40].

Based on the electrode material and actuator design, various electrode decomposition process is possible including spray coating, stamping, inkjet printing, application with paintbrush and multistep methods [37]. Critical factors influential on the DEA actuator are good adhesion, uniform thickness, and electrode particle distribution [40]. Figure 1.9 compares the properties of some electrodes.

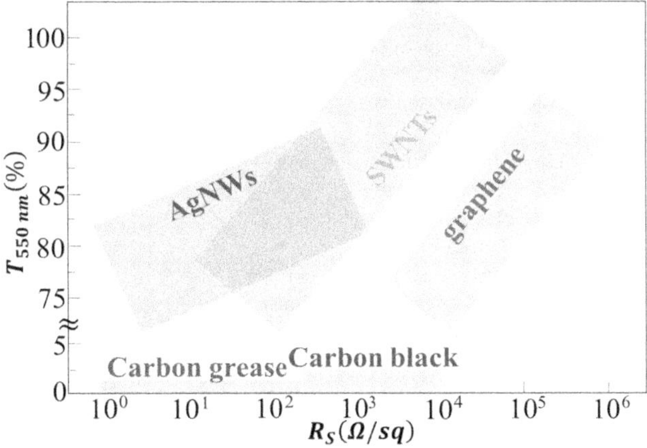


**FIGURE 1.9** Transmittance and sheet resistance of some common electrodes [39].

### 1.2.3 Dielectric Elastomers

As long as, providing requirements including low electrical conductivity, high electric permittivity and considerable elongation at break, any elastomer can be used as the elastomeric film [37]. However, the most frequently reported elastomers to fabricate DEAs are polyacrylics, silicones, and polyurethanes [36, 38].

VHB4905 and VHB4910 are the main acrylic elastomers to create DEAs sufficient for general applications and observations [38]. VHB films have a high dielectric permittivity (3.21 at 1 kHz) and linear strain (up to 600%). Significant viscoelastic loss and temperature sensitivity limit the application of VHB films and all acrylic DEAs.

Polyurethane DEAs have limited application in artificial muscle due to their low strain. Polyacrylic exhibits higher strains and lower Young's modulus and dielectric permittivity. Silicone DEAs, possessing intermediate properties, fill the gap between the two former categories. They also have the advantage of low viscoelastic losses and sustainable operation. Due to lower viscosity, silicon-based DEAs perform more reliably at higher frequencies [39]. However, their actuation voltage is higher than that of polyacrylic elastomers [37].

Table 1.1 [41] and Table 1.2 [39] summarize some of the reported properties of the common DE materials.

**TABLE 1.1**
**DE Material Properties [41]**

| | Advantages | Disadvantages |
|---|---|---|
| **Silicon** | • High electromechanical coupling efficiency<br>• Low creep<br>• Good temperature tolerance (–50 to 200 °C)<br>• Long service life<br>• Excellent frequency response performance (resonance frequency >1000 Hz)<br>• High power density fast response | • Low dielectric constant (2.8)<br>• High triggering electric field (72–350 kV/mm)<br>• Moderate actuated deformation (5.6–100%) |
| **Acrylics** | • Large driving deformation under pre-stretching<br>• High elongation at break (>600%)<br>• High dielectric constant (4.7)<br>• Large driving deformation (>100%)<br>• High energy density (3.4 MJ/m$^3$) | • Severe hysteresis loss<br>• Long drive response time<br>• Poor high-frequency response performance |
| **Polyurethane** | • Higher dielectric constant (6,7)<br>• Large actuated output force (0.14–1.6 MPa) | • High elastic modulus (>1 MPa)<br>• Low breakdown field strength (25 kV/mm)<br>• Small actuated deformation (<5%) |

**TABLE 1.2**
**Mechanical Properties of Common DE Material [39]**

| References | Type | Pre-Strain $(x, y)$ (%) | Maximum Strain (%) | Work Density per Cycle (J/cm$^3$) | Maximum Energy Efficiency |
|---|---|---|---|---|---|
| Mishra et al. [3], Mirfakhrai et al. [42], Kornbluh et al. [43] | Dielectric elastomer (VHB 4910) | 300, 300 | 158 (Circular area strain) | <3.4 (0.15 typical) | 80 (30 typical) |
| Mishra et al. [3], Mirfakhrai et al. [42], Kornbluh et al. [43] | Dielectric elastomer (VHB 4910) | 540, 75 | 215 (Linear area strain) | <1.36 | 80 (30 typical) |
| Tan et al. [44], Pelrine et al. [45] | Dielectric elastomer (polyurethane acrylate) | 0,0 | 66.7 (Circular area strain) | 0.087 | 80 |
| Pelrine et al. [45] | Dielectric elastomer (Sylgard 186) | 0,0 | 32 (Circular area strain) | 0.082 | 90 |
| Kornbluh et al. [43], Brochu and Pei [46] | Dielectric elastomer (CF19-2186) | 45, 45 | 64 (Circular area strain) | 0.75 | 79 |
| Kornbluh et al. [43] | Dielectric elastomer (CF19-2186) | 100, 0 | 63 (Linear area strain) | 0.2 | - |
| Kornbluh et al. [43] | Dielectric elastomer (HS3) | 280,0 | 117 (Linear area strain) | 0.16 | - |

### 1.2.4 DE Actuator Configurations

Thanks to the simple working mechanism and broad compatibility of dielectric elastomer actuators, various configurations have been suggested using different geometries and boundary conditions [38]. All the configurations incorporate a planar dielectric elastomer film coated with thin layers of electrodes. However, each of them provides single or multi degrees of freedom based on their geometries, mechanical and electrical boundary conditions. The widely used configurations are: planar, stacked, spring roll, and balloon [35, 36]. These configurations are shown in Figure 1.10.

Planar actuator obeys the fundamental working mechanism illustrated as Figure 1.7. Therefore, planar actuators provide in-plane expansion or thickness reduction. Stacking some DE films creates multi-layered actuators. Spider actuators also work based on the planar expansion. In a spider actuator, the edge of a circular DE film is attached to a radial mechanism.

In some cases, when deformation in one particular direction is required, introducing some anisotropy would provide deformation along the preferred direction. For instance, fixing the deformation along the two opposite edges of the rectangular planar actuator, forces it deform along the perpendicular direction in the plane. Moreover, application of an inactive layer, attached to one of the DE actuator surfaces, constrains the expansion in the fixed plane while the other surface expands freely. Consequently, the actuator experiences out-of-plane deformation. Constraining one edge of this actuator creates a bending deformation.

Rolling the DE actuator, basically possible because of elastomer's softness, to create a tube makes an actuator to generate especial deformations. Based on the particular need, the final rolled actuator may have one (tubular actuator) or more

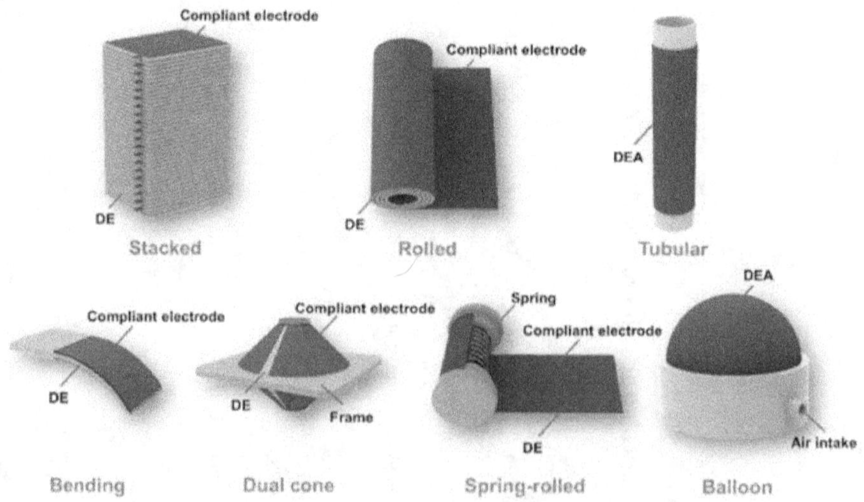


**FIGURE 1.10** Dielectric elastomer actuator configurations [41].

layers (Rolled actuators. The position of the electrodes and electric field direction determine degrees of freedom and type of deformation. This configuration is similar to actual muscles.

Other interesting configurations are conical ones. Fixing the peripheral edge of an electrode circular DE film by a frame makes the film inflate as actuated by an external electric field. The final configuration is similar to a cone. Two conical actuators that are put in front of each other create dual cone actuators.

Making use of other mechanical and electrical elements, in addition to simple working principle of DEs, creates more complicate systems with particular potentials. For instance, spring roll actuators use a rolling DEA around a core spring.

### 1.2.5 Electromechanical Instability

Since the actuation voltage of the dielectric elastomer is high, a significant challenge when using DEAs is electromechanical instability (EMI). As the thickness of DE film reduces, the electric field induced by the applied voltage increases. This creates a positive feedback loop that leads to more thickness reduction until the film fails because of electromechanical instability [36]. This phenomenon, known as pull-in instability, happens when wrinkling patterns appear on the surface of the elastomer film since the Maxwell stress exceeds the third principal stress.

Conducted researches [47] have introduced hessian criterion to quantify the stretch threshold for pull-in instability. Appropriate electromechanical control [47] and loading protocols [48] are observed to have tuning role when it comes to electromechanically instability.

Pre-stretch is proved to positively impact on the EMI and provide safer actuation in DEAs. However, maintaining the tension on the pre-stretched film limits the DEA applications [36].

Other ways to suppress EMI of DEAs are to synthesize an interpenetrating polymer network (IPNs) and UV-curable DE (UV-DE) with a tuned stress–strain relationship (bimodal-network DEs) [36].

Other failure modes of dielectric elastomers happen based on material and dielectric strength [49].

### 1.2.6 DEA Applications

So many applications have been reported for dielectric elastomer actuators since their emergence: controllable valves, pumps, loudspeakers, optical lenses, grippers, soft robots [38], and rehabilitation devices [36]. A comprehensive description of DEA actuators' applications is presented in Table 1.3 [40].

### 1.2.7 Dielectric Elastomer Sensor

Dielectric elastomers can work in the opposite direction and transform mechanical energy to electrical one, as well. In this application, as the elastomer film

**TABLE 1.3**
**Applications of DEAs [40]**

| Application Field Robotics | Description |
|---|---|
| Medical devices | Dielectric elastomeric actuators can be used to power the movement of robotic joints, allowing for precise and smooth movement. |
| Automotive | These actuators can be used in medical devices such as prosthetic limbs and artificial heart valves, enabling precise and controlled movement. |
| Industrial automation | Dielectric elastomeric actuators can be used in automotive applications such as suspension systems, power steering, and braking systems, providing precise and efficient control. |
| Aerospace | These actuators can be used in industrial automation systems, such as in conveyor belts and packaging machines, to provide precise and consistent movement. |
| Consumer electronics | Dielectric elastomeric actuators can be used in aerospace applications such as aircraft wing flaps and control surfaces, providing precise and efficient movement. |
| Building automation | These actuators can be used in consumer electronics such as smartphones and tablets, providing haptic feedback and tactile sensation. |
| Wearable technology | Dielectric elastomeric actuators can be used in building automation systems such as door openers, window shades, and ventilation systems, providing precise and efficient control. |
| Musical instruments | These actuators can be used in wearable technology such as fitness trackers, smart watches, and augmented reality devices, providing precise and consistent movement. |
| Biomedical research | Dielectric elastomeric actuators can be used in musical instruments such as electronic pianos and guitars, providing precise and consistent movement for accurate sound production. |
| Application field Robotics | These actuators can be used in biomedical research, such as in the study of muscle movement and tissue engineering, to provide precise and controlled movement. |

deforms because of mechanical force exertion, the electric capacitance of the film changes. Changes in the electric capacitance represent applied mechanical force and deformation [50]. Figure 1.11 presents the working principle of a DES.

These sensors are reported to sense mechanical deformation [51] including strain [52], pressure [53] and shear [54], as well as force [55].

### 1.2.8 Dielectric Elastomer Generator

In addition to actuator and sensor, dielectric elastomers are able to act as generators. Energy conversion in Dielectric Elastomer Generators (DEGs) happens in four steps: (1) The elastomer has the minimum capacitance in its initial state. Then the external mechanical force is exerted and makes the elastomer expand and increases the capacitance to the maximum level. (2) A power source

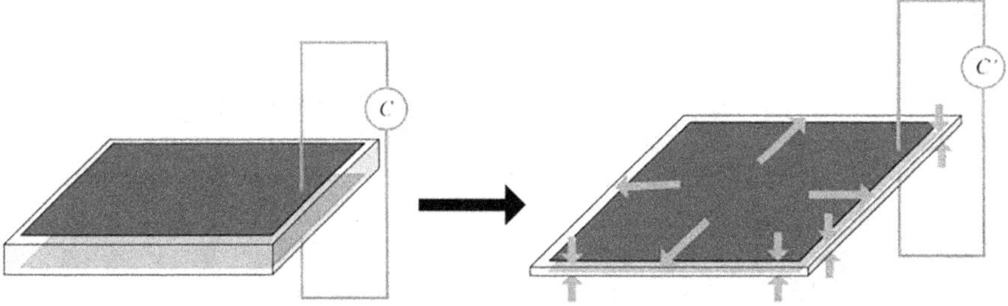


**FIGURE 1.11** Working principle of a dielectric elastomer sensor.

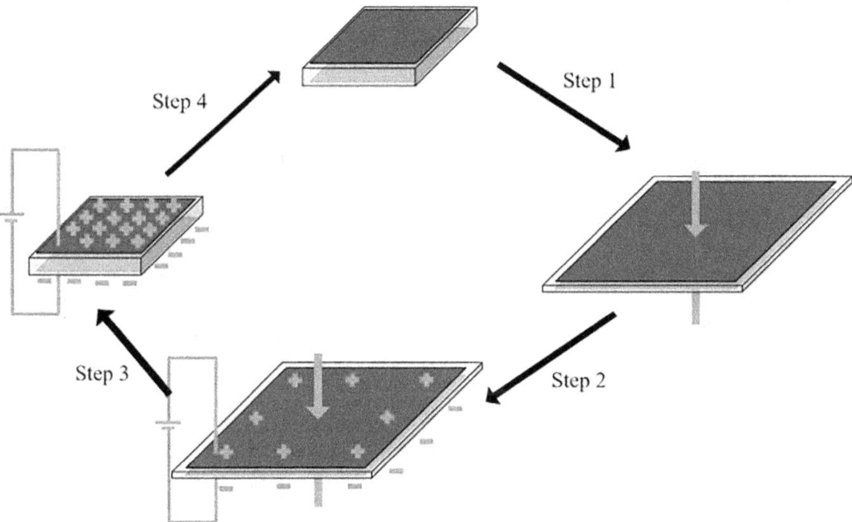


**FIGURE 1.12** Energy harvesting cycle of dielectric elastomer generator.

distributes charges on the electrodes, and then the source is eliminated. The charges remain constant. (3) The mechanical forces are removed and the film transforms to its relaxed configuration. Charges travel increase the electrical energy in DEG. (4) The gained electrical energy is collected. In summary, as the charged dielectric elastomer generator moves toward its refence configuration electrical energy is generated [56]. This is briefly shown in Figure 1.12.

The performance of a DEG is evaluated using energy density and electromechanical energy conversion efficiency. Energy density is the ratio of energy generated per mass unit of the material. The ratio of the generated energy in the last step of a cycle to the mechanical energy of the cycle.

Described energy conversion cycle provide an excellent opportunity to harvest energy from different sources. The main sources for mechanical force to be used in DEGs are wind energy [57], wave energy [58] and human motion energy [59].

## 1.3 CONCLUSION

This chapter reviews the fundamental concepts on dielectric elastomer actuator as an important class of electroactive polymers. Introduction on the working principle, main parts, properties and possible configurations clarified potential applications and challenges of dielectric elastomer actuators. Though the book is mainly focused on the DE actuators, brief introductions on the DE sensors and generators were presented in order to encourage the readers to consider other possibilities provided by dielectric elastomer materials. Part one is continued by an investigation of the anisotropic dielectric elastomer actuators.

## REFERENCES

[1] R. Samatham, K. Kim, D. Dogruer, H. Choi, M. Konyo, J. Madden, Y. Nakabo, J.-D. Nam, J. Su, S. Tadokoro, *Active polymers: An overview, Electroactive polymers for robotic applications: Artificialmuscles and sensors*, (2007) 1–36.

[2] A.V. Maksimkin, T. Dayyoub, D.V. Telyshev, A.Y. Gerasimenko, Electroactive polymer-based composites for artificial muscle-like actuators: A review, *Nanomaterials*, 12(13) (2022) 2272.

[3] S. Mishra, L. Unnikrishnan, S.K. Nayak, S. Mohanty, Advances in piezoelectric polymer composites for energy harvesting applications: A systematic review, *Macromolecular Materials and Engineering*, 304(1) (2019) 1800463.

[4] R.A. Surmenev, T. Orlova, R.V. Chernozem, A.A. Ivanova, A. Bartasyte, S. Mathur, M.A. Surmeneva, Hybrid lead-free polymer-based nanocomposites with improved piezoelectric response for biomedical energy-harvesting applications: A review, *Nano Energy*, 62 (2019) 475–506.

[5] M. Khan, L. Tiehu, A. Khurram, Active sites determination and de-aggregation of detonation nanodiamond powder, *Chaing Mai Journal of Science*, 44 (2017) 114.

[6] A. Hayat, N. Shaishta, S.K.B. Mane, J. Khan, A. Hayat, Rational Ionothermal copolymerization of TCNQ with PCN semiconductor for enhanced Photocatalytic full water splitting, *ACS Applied Materials & Interfaces*, 11(50) (2019) 46756–46766.

[7] L.J. Romasanta, M.A. López-Manchado, R. Verdejo, Increasing the performance of dielectric elastomer actuators: A review from the materials perspective, *Progress in Polymer Science*, 51 (2015) 188–211.

[8] M. Park, J. Kim, H. Song, S. Kim, M. Jeon, Fast and stable ionic electroactive polymer actuators with PEDOT: PSS/(Graphene–Ag-Nanowires) nanocomposite electrodes, *Sensors*, 18(9) (2018) 3126.

[9] L. Chang, Y. Liu, Q. Yang, L. Yu, J. Liu, Z. Zhu, P. Lu, Y. Wu, Y. Hu, Ionic electroactive polymers used in bionic robots: A review, *Journal of Bionic Engineering*, 15 (2018) 765–782.

[10] Y. Bar-Cohen, V. Cardoso, C. Ribeiro, S. Lanceros-Méndez, Electroactive polymers as actuators, *Advanced Piezoelectric Materials*, (2017) 319–352. Woodhead Publishing in Materials.

[11] M. Shahinpoor, K.J. Kim, Ionic polymer-metal composites: I. *Fundamentals, Smart Materials and Structures*, 10(4) (2001) 819.

[12] R. Baughman, Conducting polymer artificial muscles, *Synthetic Metals*, 78(3) (1996) 339–353.

[13] M. Fuchiwaki, K. Tanaka, K. Kaneto, Planate conducting polymer actuator based on polypyrrole and its application, *Sensors and Actuators A: Physical*, 150(2) (2009) 272–276.
[14] V.V. Tran, S. Lee, D. Lee, T.-H. Le, Recent developments and implementations of conductive polymer-based flexible devices in sensing applications, *Polymers*, 14(18) (2022) 3730.
[15] R. Ravichandran, S. Sundarrajan, J.R. Venugopal, S. Mukherjee, S. Ramakrishna, Applications of conducting polymers and their issues in biomedical engineering, *Journal of the Royal Society Interface*, 7(suppl_5) (2010) S559–S579.
[16] G. Spinks, G. Wallace, R. Baughman, L. Dai, *Carbon Nanotube Actuators: Synthesis, Properties, and Performance, Electroactive Polymers (EAP) Actuators as Artificial Muscles, Reality, Potential and Challenges*, (2004) 261–295. SPIE.
[17] V.N. Popov, Carbon nanotubes: Properties and application, *Materials Science & Engineering R: Reports*, 43(3) (2004) 61–102.
[18] P.M. Ajayan, O.Z. Zhou, Applications of carbon nanotubes, *Carbon nanotubes: Synthesis, structure, properties, and applications*, (2001) 391–425. Springer.
[19] H. Dai, Carbon nanotubes: Opportunities and challenges, *Surface Science*, 500(1–3) (2002) 218–241.
[20] H. Wang, Z. Wang, J. Yang, C. Xu, Q. Zhang, Z. Peng, Ionic gels and their applications in stretchable electronics, *Macromolecular Rapid Communications*, 39(16) (2018) 1800246.
[21] Z. Liu, P. Calvert, Multilayer hydrogels as muscle-like actuators, *Advanced Materials*, 12(4) (2000) 288–291.
[22] H. Xu, C. Han, X. Liu, Z. Li, J. Liu, Z. Sun, A highly flexible and stretchable ionic artificial muscle, *Sensors and Actuators A: Physical*, 332 (2021) 113190.
[23] C. Zhang, B. He, Z. Wang, Y. Zhou, A. Ming, Application and analysis of an ionic liquid gel in a soft robot, *Advances in Materials Science and Engineering*, 2019 (2019).
[24] Z. Liu, Y. Wang, Y. Ren, G. Jin, C. Zhang, W. Chen, F. Yan, Poly (ionic liquid) hydrogel-based anti-freezing ionic skin for a soft robotic gripper, *Materials Horizons*, 7(3) (2020) 919–927.
[25] Y. Qiu, K. Park, Environment-sensitive hydrogels for drug delivery, *Advanced Drug Delivery Reviews*, 53(3) (2001) 321–339.
[26] T. Wang, M. Farajollahi, Y.S. Choi, I.-T. Lin, J.E. Marshall, N.M. Thompson, S. Kar-Narayan, J.D. Madden, S.K. Smoukov, Electroactive polymers for sensing, *Interface focus*, 6(4) (2016) 20160026.
[27] Q. Li, Q. Wang, Ferroelectric polymers and their energy-related applications, *Macromolecular Chemistry and Physics*, 217(11) (2016) 1228–1244.
[28] Y. Bar-Cohen, Electroactive polymers as an enabling materials technology, *Proceedings of the Institution of Mechanical Engineers, Part G: Journal of Aerospace Engineering*, 221(4) (2007) 553–564.
[29] H. Li, R. Wang, S.T. Han, Y. Zhou, Ferroelectric polymers for non-volatile memory devices: A review, *Polymer International*, 69(6) (2020) 533–544.
[30] K.L. Kim, W. Lee, S.K. Hwang, S.H. Joo, S.M. Cho, G. Song, S.H. Cho, B. Jeong, I. Hwang, J.-H. Ahn, Epitaxial growth of thin ferroelectric polymer films on graphene layer for fully transparent and flexible nonvolatile memory, *Nano Letters*, 16(1) (2016) 334–340.
[31] S.W. Ula, N.A. Traugutt, R.H. Volpe, R.R. Patel, K. Yu, C.M. Yakacki, Liquid crystal elastomers: An introduction and review of emerging technologies, *Liquid Crystals Reviews*, 6(1) (2018) 78–107.

[32] J. Wu, Y. Xiao, Y. Zhang, D. Fang, A constitutive model of liquid crystal elastomers with loading-history dependence, *Journal of the Mechanics and Physics of Solids*, 174 (2023) 105258.

[33] Y. Wang, C. Sun, E. Zhou, J. Su, Deformation mechanisms of electrostrictive graft elastomer, *Smart Materials and Structures*, 13(6) (2004) 1407.

[34] H. Grellmann, F.M. Lohse, V.G. Kamble, H. Winger, A. Nocke, R. Hickmann, S. Wießner, C. Cherif, Fundamentals and working mechanisms of artificial muscles with textile application in the loop, *Smart Materials and Structures*, 31(2) (2021) 023001.

[35] G.-Y. Gu, J. Zhu, L.-M. Zhu, X. Zhu, A survey on dielectric elastomer actuators for soft robots, *Bioinspiration & Biomimetics*, 12(1) (2017) 011003.

[36] Y. Guo, Q. Qin, Z. Han, R. Plamthottam, M. Possinger, Q. Pei, Dielectric elastomer artificial muscle materials advancement and soft robotic applications, *SmartMat*, 4(4) (2023) e1203.

[37] I.V.E. Bezsudnov, A.G. Khmelnitskaia, A.A. Kalinina, S.A.E. Ponomarenko, Dielectric elastomer actuators: Materials and design, *Uspehi himii*, 92(2) (2023) 1–44.

[38] H. Böse, J. Ehrlich, Dielectric elastomer sensors with advanced designs and their applications, in: *Actuators*, MDPI, 2023, pp. 115.

[39] J.-H. Youn, S.M. Jeong, G. Hwang, H. Kim, K. Hyeon, J. Park, K.-U. Kyung, Dielectric elastomer actuator for soft robotics applications and challenges, *Applied Sciences*, 10(2) (2020) 640.

[40] J.N. Suresh, I. Arief, K. Naskar, G. Heinrich, M. Tahir, S. Wießner, A. Das, The role of chemical microstructures and compositions on the actuation performance of dielectric elastomers: A materials research perspective, *Nano Select*, 4(5) (2023), 289–315.

[41] Y. Wang, X. Ma, Y. Jiang, W. Zang, P. Cao, M. Tian, N. Ning, L. Zhang, Dielectric elastomer actuators for artificial muscles: A comprehensive review of soft robot explorations, *Resources Chemicals and Materials*, 1 (2022) 308–324.

[42] T. Mirfakhrai, J.D. Madden, R.H. Baughman, Polymer artificial muscles, *Materials Today*, 10(4) (2007) 30–38.

[43] R.D. Kornbluh, R. Pelrine, Q. Pei, S. Oh, J. Joseph, Ultrahigh strain response of field-actuated elastomeric polymers, in: *Smart Structures and Materials 2000: Electroactive Polymer Actuators and Devices (Eapad)*, SPIE, 2000, pp. 51–64.

[44] M.W.M. Tan, G. Thangavel, P.S. Lee, Enhancing dynamic actuation performance of dielectric elastomer actuators by tuning viscoelastic effects with polar crosslinking, *NPG Asia Materials*, 11(1) (2019) 62.

[45] R. Pelrine, R. Kornbluh, J. Joseph, R. Heydt, Q. Pei, S. Chiba, High-field deformation of elastomeric dielectrics for actuators, *Materials Science and Engineering: C*, 11(2) (2000) 89–100.

[46] P. Brochu, Q. Pei, Advances in dielectric elastomers for actuators and artificial muscles, *Macromolecular Rapid Communications*, 31(1) (2010) 10–36.

[47] A. Khurana, M. Joglekar, G. Zurlo, Electromechanical stability of wrinkled dielectric elastomers, *International Journal of Solids and Structures*, 246 (2022) 111613.

[48] Y. Su, W. Chen, M. Destrade, Tuning the pull-in instability of soft dielectric elastomers through loading protocols, *International Journal of Non-Linear Mechanics*, 113 (2019) 62–66.

[49] J.-S. Plante, S. Dubowsky, Large-scale failure modes of dielectric elastomer actuators, *International Journal of Solids and Structures*, 43(25–26) (2006) 7727–7751.

[50] N. Ni, L. Zhang, Dielectric elastomer sensors, in: *Elastomers*, IntechOpen Rijeka, 2017, pp. 231–253.
[51] M. Rosenthal, N. Bonwit, C. Duncheon, J. Heim, Applications of dielectric elastomer EPAM sensors, in: *Electroactive Polymer Actuators and Devices (EAPAD) 2007*, SPIE, 2007, pp. 410–416.
[52] B. Huang, M. Li, T. Mei, D. McCoul, S. Qin, Z. Zhao, J. Zhao, Wearable stretch sensors for motion measurement of the wrist joint based on dielectric elastomers, *Sensors*, 17(12) (2017) 2708.
[53] H. Böse, D. Ocak, J. Ehrlich, Applications of pressure-sensitive dielectric elastomer sensors, in: *Electroactive Polymer Actuators and Devices (EAPAD) 2016*, SPIE, 2016, pp. 451–463.
[54] H. Zhang, M.Y. Wang, Multi-axis soft sensors based on dielectric elastomer, *Soft Robotics*, 3(1) (2016) 3–12.
[55] J. Cheng, Z. Jia, T. Li, Dielectric-elastomer-based capacitive force sensing with tunable and enhanced sensitivity, *Extreme Mechanics Letters*, 21 (2018) 49–56.
[56] K. Di, K. Bao, H. Chen, X. Xie, J. Tan, Y. Shao, Y. Li, W. Xia, Z. Xu, E. Shiju, Dielectric elastomer generator for electromechanical energy conversion: A mini review, *Sustainability*, 13(17) (2021) 9881.
[57] C. Zhang, Z. Lai, M. Li, D. Yurchenko, Wind energy harvesting from a conventional turbine structure with an embedded vibro-impact dielectric elastomer generator, *Journal of Sound and Vibration*, 487 (2020) 115616.
[58] G. Moretti, G.P.R. Papini, M. Righi, D. Forehand, D. Ingram, R. Vertechy, M. Fontana, Resonant wave energy harvester based on dielectric elastomer generator, *Smart Materials and Structures*, 27(3) (2018) 035015.
[59] K. Ylli, D. Hoffmann, A. Willmann, P. Becker, B. Folkmer, Y. Manoli, Energy harvesting from human motion: Exploiting swing and shock excitations, *Smart Materials and Structures*, 24(2) (2015) 025029.

# 2 Anisotropic Dielectric Elastomer Composites

## 2.1 LAYERED DIELECTRIC ELASTOMERS

Combining Equations (2.1) and (2.2) introduces an equation for the strain induced in a dielectric elastomer actuator:

$$s_z = \frac{p}{Y} = \frac{1}{Y}\varepsilon_0\varepsilon_r\mathbb{E}^2 \tag{2.1}$$

where $\varepsilon_r$, $\varepsilon_0$, $\mathbb{E}$ and $Y$ are relative permittivity, vacuum permittivity, external electric field and Young's modulus of the elastomer film, respectively. To increase the strain, one needs to increase relative permittivity or electric field or decrease the Young's modulus. Since changes of the electric field affect the strain quadratically, it is more efficient to increase this factor, rather than the other ones.

The electric field caused by a potential difference obeys the following equation:

$$\mathbb{E} = \frac{\mathbb{V}}{d} \tag{2.2}$$

where $\mathbb{V}$ and $d$, are potential difference and the elastomer film thickness. To increase the electric field, the voltage should grow, or the film thickness should decrease.

However, when it comes to produce large deformation, a vital disadvantage that limits application of dielectric elastomer actuators is their high actuation voltage. In the other word, voltages needed to actuate a dielectric elastomer are high enough already and it is not safe to increase it any more. Higher voltages may lead to dielectric breakdown and make the actuator fail. Therefore, attention has been attracted to other potential solutions to enhance the actuation strain and efficiency of dielectric elastomer actuators. A possible method is to reduce effective thickness utilizing stack or multi-layered actuators.

According to Equation (1.2), if the needed electric field to actuate a dielectric elastomer film with thickness of $d$ is E, reducing the thickness to $d/N$, decrease the actuation voltage to $V/N$, as well.

Hence, multi-layered DEAs, composed of layers connected electrically parallel, benefit from lower actuation voltages compared to a single-layer DEA with the same total thickness. Consequently, increasing the number of layers in constant thickness leverages the actuation strain. Multi-layered actuators are also reported to generate higher output forces [1, 2].


DOI: 10.1201/9781003571469-3

Layering also makes it possible to induce an inhomogeneous internal electric field using electrodes with different geometries. The presence of adjacent layers also prevents local instabilities [3].

To fabricate multi-layered DEA, some layers of elastomer films covered by electrodes are piled up. The layers are mechanically connected in series and the strain of all layers sum up to create the multi-layered actuation strain. Multi-layered actuators can be produced in various sizes and configurations. Figure 2.1 shows the schematic representation of a planar multi-layered DE actuator.

According to the direction of induced deformation, multi-layered actuators are categorized into two main groups [4]: Linear Contraction DEA and Linear Expansion DEA.

The first multi-layered actuator group is used to create deformation along the elastomer film thickness, based on the actuator's thickness changes, $z$ direction in Figure 2.2. In this group, planar deformation is low. Limiting the deformation in one direction maximizes the deformation along the other. Most of the actuator in this group are composed of high number of layers, that is no surprise considering the direction of actuation in interest. The final contractive deformation of the multi-layered is the sum of compressions in all layers.

On the other hand, there are expanding actuators. These actuators provide the need for actuation along the area of the elastomer film: length or width or both, $x$ and $y$ directions in Figure 2.3.

Despite vast potentials, development and application of multi-layered actuators suffer from a slow peace. The main factors to blame are technical difficulties of fabrication process that cannot produce multi-layered actuators with required qualities [5]. For example, layers must bond strongly to ensure the structural integrity of the multi-layer. In this regard, fabrication processes, their cons and pros, and fabrication innovations have been the subject of some research on multi-layered dielectric elastomer actuators [5–19]. Some of these investigations have specifically targeted improved lifespan of multi-layered DEAs [20].

Multi-layered actuators face other challenges, as well. First, decreasing the thickness of elastomer films is limited by the thickness of the electrodes. More thickness reduction leads to lower actuation strains because of the

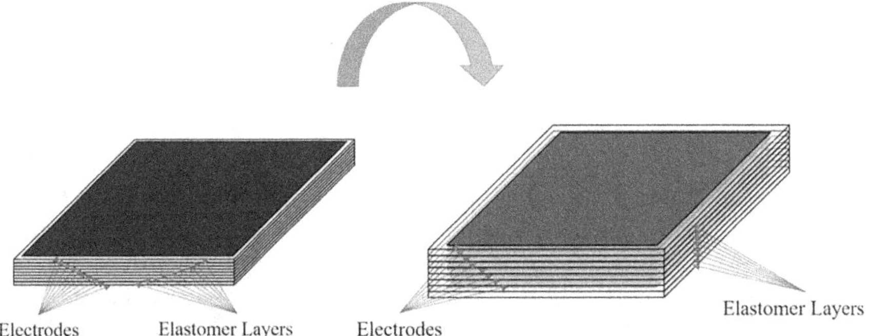


**FIGURE 2.1** Dielectric elastomer multi-layered actuator.

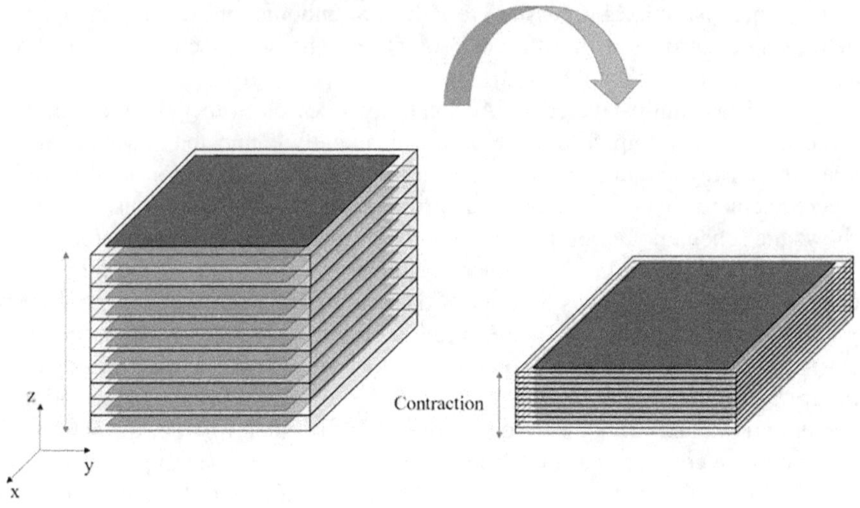


FIGURE 2.2 Multi-layered linear contraction DEA.

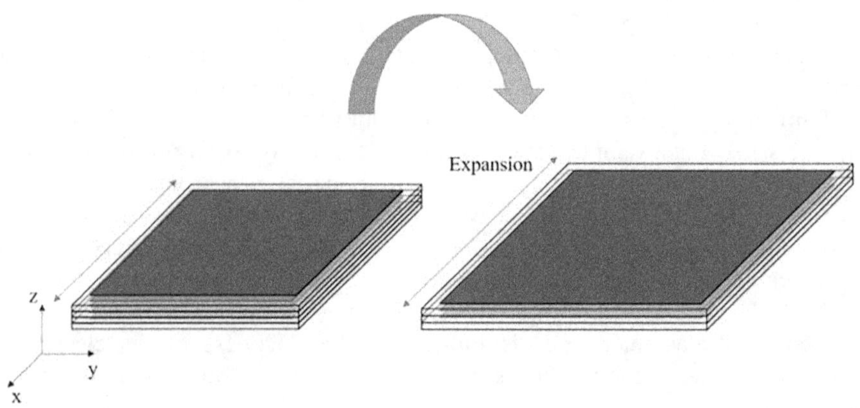


FIGURE 2.3 Multi-layered linear expansion DEA.

elastomers-electrodes thickness ratio and the structure stiffness. The elastomers-electrodes thickness ratio can also affect the adhesion between electrodes and elastomers and may cause delamination. In ideal configuration, thickness of the electrodes is negligible, compared to the elastomer films thickness [21]. Another issue with stacked and multi-layered DEA is their susceptibility to electro-thermal breakdown [22–24].

It was mentioned that stacked actuators are made of a high number of layers. Increased number of layers harden the fabrication process. A possible method to address this issue is folded actuators [4]. The fabrication process of folded actuators is more straightforward because they are made of a monolithic film coated by

**FIGURE 2.4** Folded DEA.

one layer of electrode [25]. Moreover, it functionally is equivalent to usual multi-layered actuators and offers similar advantages [26]. Figure 2.4 shows the schematic view of a folded actuator.

Rolled actuators are manufactured based on the similar process except the strip rolls. Application of electric field induces radial Maxwell-stress in the actuator [27]. Figure 2.5 presents fabrication process of a rolled dielectric elastomer actuator used in a tactile displace [28].

On top of the introduced configurations, more innovative structures are methods are proposed for the dielectric elastomer actuators [29–31]. Unique geometries and features of these newly designed actuators are valuable tools to devise more efficient devices.

Including inactive layers with particular mechanical properties gives special features and potentials for multi-layered actuators. For example, a substrate, which is stiff in tension and flexible in bending, creates unimorphs and bimorphs bending actuators [21, 32–35].

A promising opportunity due to the layering concept is improved dielectric and mechanical properties provided by controlling the composition and layer structure of the actuator [36].

Proposed applications for stack and multi-layered DEAs are active suspension and vibration control [37–40], biomimetic legged robots [40], crawling robots [41], jumping robots [1], soft gripers [2, 42], cardiac assist devices [43–46], impedance pump [47, 48], jellyfish-inspired robot [49], wearable haptic display [50, 51], steerable tip [52], flapping-wing robot [53], rotary motor [54], Skin-Mountable Vibrotactile Stimulator [55] and flapping fins [56].

Various electromechanical models have been proposed to investigate static and dynamic behavior of a multi-layered dielectric elastomer actuator and how parameters including actuating voltage, layers number and thickness affect it [23, 24, 57–60].

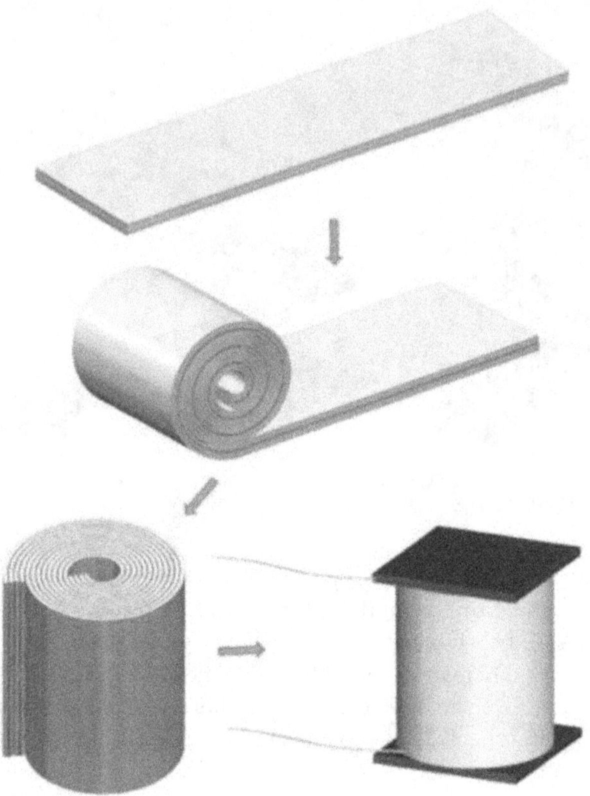

**FIGURE 2.5** Rolled DEA [28].

## 2.2 FIBER-REINFORCED DIELECTRIC ELASTOMERS

According to the theory proposed by Zhao and Suo [61], an elastomer with compliant behavior in small stretches that gets stiffer at modest stretch, is likely to eliminate or survive the electromechanical instability and reach large actuation deformation. An approach to achieve such a form in stress-strain curve is to reinforced the elastomer with stiffer nature. Here, we consider fiber-reinforced elastomers, mainly.

An elastomer, reinforced by a family of fibers, as depicted in Figure 2.6, shows different properties along the different axes. If the fibers are stiffer than matrix elastomer, it particularly is stiffer against the loading along the fiber orientation. Therefore, the application of an electric field upon a DEA induces anisotropic deformation in it. The presence of fibers limits the deformation in their direction, so the mechanical energy is consumed to create deformation in other directions. On the other hand, when the induced deformation is not large enough to create tension in fibers, the soft nature of the elastomer dictates the stress-strain curve of

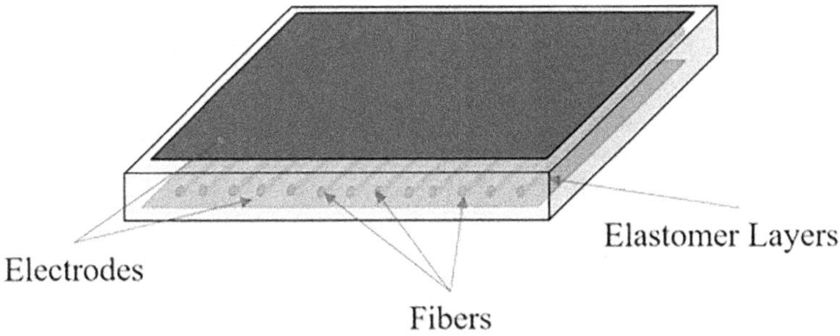


**FIGURE 2.6** Fiber-reinforced dielectric elastomer actuator.

the actuator. As the deformation grows, fibers experience tension. They constrained strain and causes stiffer behavior.

The idea of reinforcing electroactive polymers with infusion phase including nanotube [62, 63], ceramic [63, 64], carbon black particles [65, 66], polyaniline or ceramic powder fillers [67], titanium dioxide particles [66], rigid conducting inclusions [68], silver nanoparticle [69] and Al particle [70], was followed by early research on inhomogeneous dielectric elastomers that have examined the effect of fillers addition on actuation stress, elastic energy density and strain [71]. These effects are interpreted based on particle-matrix interaction.

Soon, cylindrical dielectric elastomer actuators reinforced with fibers attracted researchers' attention. The initial idea was suggested by Goulbourne based on McKibben actuators [72]. He proposed to devise a family of helical, flexible, and inextensible fibers on the outer surface of the cylinder. This configuration not only benefits from the simultaneous axial shortening and radial expansion but also is reinforced by the fiber family due to improved load-bearing capacity [73]. Combined nonlinear elasticity theory and Maxwell stress have been used to explain how the actuator behaves [72]. The results revealed the dependency of electromechanical behavior of the fiber-reinforced cylindrical actuator on the fiber family orientation and applied voltage [74]. Fiber reinforcement has also increased output force. According to the numerical results, force and axial deformation have a bilinear relation [73].

Cylindrical dielectric elastomer actuators reinforced with one and two families of fibers are introduced to perform as torsional actuators [75, 76]. Constraints caused by helical inextensible fibers make the actuator rotates as actuated by radial voltage. Effect of material parameters, fibers arrangement, external force and torque on the uniform and nonuniform torsion of the actuator are also investigated using numerical and analytical models [75–77]. The model also facilitates snap-through instability investigations.

In microstructure scale, according to the conducted studies on dielectric elastomer composites [78], electromechanical coupling is amplified by combination of constituent materials or polydisperse microstructures. The numerical solution also demonstrated enhanced stability by considering anisotropy parameters [79].

The works led to an approximate solution for the overall elastic dielectric response of elastomers in the classical limit of small deformations and moderate electric fields [80].

Motivated by experimental observations, and the mentioned theory [61], studies on the electromechanical behavior of dielectric elastomer actuators have also led to some models to capture the effect of reinforcing fibers, too. The resultant findings confirmed the positive impact of fibers on the electromechanical stability of these actuators. The models have examined the impact of fibers features including stiffness, distribution and spacing on the stability of a DE membrane [81]. Some electromechanical models have evolved to develop computationally efficient staggered solution algorithms that are used to solve more complex cases. These algorithms are implemented in finite element analysis that emphasizes the favorable influence of fiber reinforcements on the actuator performance [82–85]. Fibers properties including stiffness and loading are also observed to enhance mechanical and electro actuation behavior of the DEAs [86, 87].

On top of static response and behavior [84, 88], suggested electromechanical models are also applied to predict dynamic and vibrational behavior of fiber-reinforced DEAs [89–92] including phase, dynamic response, resonance frequency and how the fibers parameters affect them. In order to have the most similar behavior to the natural muscles, a more detailed investigation has modeled fibers' extensibility in addition to their arrangement [93].

Though complexity of a fiber-reinforced dielectric elastomer actuators makes its analytical investigation hard, some aspect of DEAs' actuation has been explored analytically. Solving bifurcation equations of a planar DEA has shown the effect of material anisotropy on the unstable region. According to the results, though introducing fiber anisotropy increases unstable regions, it also elevates the critical stretches. This effect is observed to be more vital for thin plates [94].

Since reinforcement mainly aims for more stable performance, some of the devised models demonstrated the electric field and deformations enhancement on the onset of instability because of embedded fibers [95]. According to the obtained results, higher volume fraction of fibers, ensures higher tolerable electric field before the instability imitation [96]. More complicated models have also simulated the effect of thermal factors on anisotropic DEAs stability via entropy [97].

On top of stability, fiber reinforcement adds unique features to the actuator like coupled bending-twisting deformation reported in planar [98] and tubular actuators [99]. Numerical and experimental results showed how tuning fiber density, orientation and stiffness adjust the actuators' behavior to meet a preferable need and provides controllability of the actuation directions.

Moreover, a DEA strengthen by stiff fibers is reported to have linear strain about 140%. The actuator was used in a bio-inspired artificial arm mimicking a natural slender muscle [100]. The other artificial arm muscle was inspired by the traditional McKibben actuators [101].The circular arrangement of the fibers enabled the actuator to mimic the spindle shape of natural muscle. The proposed model results and fabricated prototype observations revealed significant information about the actuator performance including failure modes, maximum actuation and critical voltage.

Despite the frequently-observed positive effects of fibers on the electromechanical behavior of voltage control DEAs, uniaxial tension and pure shear modes of a charge controlled planar DEA shows decreased actuation stretch, in result of fiber reinforcement in some cases. The critical factor that determines the positive or negative effect of the reinforcement are the deformation mode and geometrical size of the actuator [102].

On the other hand, a practical challenge in the field of anisotropic DEAs is the fabrication method. To address this issue some papers [103, 104] have presented experimental approaches to produce these actuators.

There are also more specific studies on fiber reinforcement considering the length of fibers. Experimental and numerical examinations [105] have shown different characteristics for DE membranes reinforced with long and short fibers and how fibers spacing influence their electromechanical behavior. These findings leaded to the design of laminates with Integrated Function of Actuating and Sensing [106].

All in all, Fiber-reinforcement has positively affected the technology of wearable actuators by providing the needed prestrain for a good performance. In this application, fibers hold the elastomer film in a prestrained state and remove the need for stiff frames. They also conduct the deformation in one particular direction [107].

Anisotropic DEAs have served to devise inchworm robots [108]. The fibers suppress bending deformation and convey the mechanical energy to generate forward motion. However, numerical solutions showed a voltage threshold needed to produce the desirable movement.

Imbedding fibers in soft DE makes a hybrid medium and provides the opportunity to except dual performance and characteristics in addition to tuning option. Using this privilege, some bio-inspired actuators are proposed and fabricated: a soft-rigid hybrid actuator made of a fiber-reinforced DEA and a skeleton backbone system inspired by the septa-muscle structure of earthworms [109], an autonomous underwater robot inspired by squid mantle made of circular membrane reinforced with intersecting fiber pattern that deforms to hemispherical geometry when inflated [110], a variable stiffness soft actuator inspired by octopus bimaculoides made of fiber-reinforced dielectric elastomer strip [111], a soft fish robot that mimics fish's tail motion made of an anisotropic convex tapered bimorph dielectric elastomer actuator [112].

## 2.3 CONCLUSION

This chapter was focused on the studied cases of fiber-reinforced and multilayered actuators dielectric elastomers. It reviewed challenges and risk, potentials, advantages and application. According to the experimental results, the actuators address concerns about the safe and reliable applications of dielectric elastomers. Motivated by the reported findings, the book is followed by the presentation of mathematical models to study different aspects of the behavior of DEAs.

## REFERENCES

[1] B. Luo, B. Li, Y. Yu, M. Yu, J. Ma, W. Yang, P. Wang, Z. Jiao, A jumping robot driven by a dielectric elastomer actuator, *Applied Sciences*, 10(7) (2020) 2241.

[2] G. Hwang, J. Park, D.S.D. Cortes, K.-U. Kyung, Mechanically strengthened electroadhesion based soft gripper with multi-layered dielectric elastomer actuator, in: *2020 3rd IEEE International Conference on Soft Robotics (RoboSoft)*, IEEE, 2020, pp. 748–753.

[3] E. Hajiesmaili, D.R. Clarke, Dielectric elastomer actuators, *Journal of Applied Physics*, 129(15) (2021). 151102.

[4] J.-H. Youn, S.M. Jeong, G. Hwang, H. Kim, K. Hyeon, J. Park, K.-U. Kyung, Dielectric elastomer actuator for soft robotics applications and challenges, *Applied Sciences*, 10(2) (2020) 640.

[5] O. Araromi, A. Conn, C. Ling, J. Rossiter, R. Vaidyanathan, S. Burgess, Spray deposited multilayered dielectric elastomer actuators, *Sensors and Actuators A: Physical*, 167(2) (2011) 459–467.

[6] X. Niu, R. Leo, D. Chen, W. Hu, Q. Pei, Multilayer stack actuator made from new prestrain-free dielectric elastomers, in: *Electroactive Polymer Actuators and Devices (EAPAD) 2013*, SPIE, 2013, pp. 429–436.

[7] F. Klug, S. Solano-Arana, H. Mößinger, F. Förster-Zügel, H.F. Schlaak, Fabrication of dielectric elastomer stack transducers (DEST) by liquid deposition modeling, in: *Electroactive Polymer Actuators and Devices (EAPAD) 2017*, SPIE, 2017, pp. 534–540.

[8] M. Corbaci, W. Walter, K. Lamkin-Kennard, Implementation of soft-lithography techniques for fabrication of Bio-Inspired multi-layer dielectric elastomer actuators with interdigitated mechanically compliant electrodes, in: *Actuators*, MDPI, 2018, pp. 73.

[9] Z. Li, M. Sheng, M. Wang, P. Dong, B. Li, H. Chen, Stacked dielectric elastomer actuator (SDEA): casting process, modeling and active vibration isolation, *Smart Materials and Structures*, 27(7) (2018) 075023.

[10] F. Klug, S. Solano-Arana, N.J. Hoffmann, H.F. Schlaak, Multilayer dielectric elastomer tubular transducers for soft robotic applications, *Smart Materials and Structures*, 28(10) (2019) 104004.

[11] E. Hajiesmaili, E. Khare, A. Chortos, J. Lewis, D.R. Clarke, Voltage-controlled morphing of dielectric elastomer circular sheets into conical surfaces, *Extreme Mechanics Letters*, 30 (2019) 100504.

[12] Z. Li, P. Dong, T. Shi, C. Tang, C. Bian, H. Chen, Elastomeric electrode and casting process for manufacturing multilayer dielectric elastomer actuators, in: *Electroactive Polymer Actuators and Devices (EAPAD) XX*, SPIE, 2018, pp. 212–221.

[13] A. Hubracht, O. Çabuk, T. Krüger, J. Maas, Fabrication and characterization of thin film-based multilayered dielectric elastomer stack transducers via submodules, in: *Electroactive Polymer Actuators and Devices (EAPAD) XXIII*, SPIE, 2021, pp. 73–84.

[14] A. Hubracht, O. Cabuk, J. Maas, 3D printing of elastomer layers for dielectric elastomer transducers, in: *Electroactive Polymer Actuators and Devices (EAPAD) XXIV*, SPIE, 2022, pp. 298–313.

[15] A.J. Cohen, M. Kollosche, M.C. Yuen, D.Y. Lee, D.R. Clarke, R.J. Wood, Batch-sprayed and stamp-transferred electrodes: a new paradigm for scalable fabrication of multilayer dielectric elastomer actuators, Advanced Functional Materials, 32(43) (2022) 2205394.

[16] P. Huang, H. Fu, M.W.M. Tan, Y. Jiang, P.S. Lee, Digital light processing 3D-printed multilayer dielectric elastomer actuator for vibrotactile device, *Advanced Materials Technologies*, 9(2) (2024) 2301642.

[17] Z. Han, Z. Peng, Y. Guo, H. Wang, R. Plamthottam, Q. Pei, Hybrid fabrication of prestrain-locked acrylic dielectric elastomer thin films and multilayer stacks, *Macromolecular Rapid Communications*, 44(15) (2023) 2300160.

[18] T.S. Krüger, O. Çabuk, J. Maas, Manufacturing process for multilayer dielectric elastomer transducers based on sheet-to-sheet lamination and contactless electrode application, in: *Actuators*, MDPI, 2023, pp. 95.

[19] W. Xiang, J. Ye, Z. Xing, P. Zhang, X. Chen, L. Wang, J. Liu, Spray-printing and spin methods to fabricate multilayered dielectric elastomer actuators embedded with liquid metal electrodes, *ACS Applied Electronic Materials*, 5(11) (2023) 6003–6013.

[20] H. Fu, Y. Jiang, J. Lv, Y. Huang, Z. Gai, Y. Liu, P.S. Lee, H. Xu, D. Wu, Multilayer dielectric elastomer with reconfigurable electrodes for artificial muscle, *Advanced Science*, 10(9) (2023) 2206094.

[21] M. Duduta, R.J. Wood, D.R. Clarke, Multilayer dielectric elastomers for fast, programmable actuation without prestretch, *Advanced Materials*, 28(36) (2016) 8058–8063.

[22] G.-K. Lau, T.-G. La, E.S.-W. Foong, M. Shrestha, Stronger multilayer acrylic dielectric elastomer actuators with silicone gel coatings, *Smart Materials and Structures*, 25(12) (2016) 125006.

[23] L.R. Christensen, O. Hassager, A.L. Skov, Electro-thermal and-mechanical model of thermal breakdown in multilayered dielectric elastomers, *AIChE Journal*, 66(8) (2020) e16275.

[24] L.R. Christensen, O. Hassager, A.L. Skov, Electro-thermal model of thermal breakdown in multilayered dielectric elastomers, *AIChE Journal*, 65(2) (2019) 859–864.

[25] F. Carpi, D. De Rossi, Contractile dielectric elastomer actuator with folded shape, in: *Smart Structures and Materials 2006: Electroactive Polymer Actuators and Devices (EAPAD)*, SPIE, 2006, pp. 99–104.

[26] F. Carpi, C. Salaris, D. De Rossi, Folded dielectric elastomer actuators, *Smart Materials and Structures*, 16(2) (2007) S300.

[27] J. Kunze, J. Prechtl, D. Bruch, B. Fasolt, S. Nalbach, P. Motzki, S. Seelecke, G. Rizzello, Design, manufacturing, and characterization of thin, core-free, rolled dielectric elastomer actuators, in: *Actuators*, MDPI, 2021, pp. 69.

[28] H. Zhao, A.M. Hussain, M. Duduta, D.M. Vogt, R.J. Wood, D.R. Clarke, Compact dielectric elastomer linear actuators, *Advanced Functional Materials*, 28(42) (2018) 1804328.

[29] J. Shintake, D. Ichige, R. Kanno, T. Nagai, K. Shimizu, Monolithic stacked dielectric elastomer actuators, *Frontiers in Robotics and AI*, 8 (2021) 714332.

[30] E. Hajiesmaili, N.M. Larson, J.A. Lewis, D.R. Clarke, Programmed shape-morphing into complex target shapes using architected dielectric elastomer actuators, *Science Advances*, 8(28) (2022) eabn9198.

[31] Y. Sun, D. Li, M. Wu, Y. Yang, J. Su, T. Wong, K. Xu, Y. Li, L. Li, X. Yu, Origami-inspired folding assembly of dielectric elastomers for programmable soft robots, *Microsystems & Nanoengineering*, 8(1) (2022) 37.

[32] O.A. Araromi, S.C. Burgess, A finite element approach for modelling multilayer unimorph dielectric elastomer actuators with inhomogeneous layer geometry, *Smart Materials and Structures*, 21(3) (2012) 032001.

[33] K. Kadooka, H. Imamura, M. Taya, Experimentally verified model of viscoelastic behavior of multilayer unimorph dielectric elastomer actuators, *Smart Materials and Structures*, 25(10) (2016) 105028.

[34] J. Ehrlich, P. Löschke, H. Böse, High-performance dielectric elastomer unimorph bending actuators without pre-stretch, in: *Electroactive Polymer Actuators and Devices (EAPAD) XXV*, SPIE, 2023, pp. 293–299.

[35] H. Böse, J. Ehrlich, P. Löschke, T. Gerlach, Modeling of the actuation performance of dielectric elastomer unimorph bending actuators consisting of different materials, in: *Electroactive Polymer Actuators and Devices (EAPAD) XXV*, SPIE, 2023, pp. 177–195.

[36] M.A. Wolak, L. Zhu, A robust multilayer dielectric elastomer actuator, *MRS Online Proceedings Library (OPL)*, 1718 (2015) 145–155.

[37] R. Karsten, P. Lotz, H.F. Schlaak, Active suspension with multilayer dielectric elastomer actuator, in: *Electroactive Polymer Actuators and Devices (EAPAD) 2011*, SPIE, 2011, pp. 755–763.

[38] Q. Xue, C. Liu, S. Xie, Y. Zhang, Y. Luo, Active vibration control of flexible thin-walled beam using multi-layer planar dielectric elastomer actuator, *Journal of Vibration and Control*, 29 (2022) 2854–2867.

[39] C. Liu, Q. Xue, S. Xie, Y. Zhang, Y. Luo, Analytical and experimental studies on active vibration control of flexible thin-walled structure using multi-layer planar dielectric elastomer actuator, *Sensors and Actuators A: Physical*, 357 (2023) 114414.

[40] C.T. Nguyen, H. Phung, T.D. Nguyen, C. Lee, U. Kim, D. Lee, H. Moon, J. Koo, H.R. Choi, A small biomimetic quadruped robot driven by multistacked dielectric elastomer actuators, *Smart Materials and Structures*, 23(6) (2014) 065005.

[41] M. Duduta, D.R. Clarke, R.J. Wood, A high speed soft robot based on dielectric elastomer actuators, in: *2017 IEEE International Conference on Robotics and Automation (ICRA)*, IEEE, 2017, pp. 4346–4351.

[42] G. Hwang, J. Park, D.S.D. Cortes, K. Hyeon, K.-U. Kyung, Electroadhesion-based high-payload soft gripper with mechanically strengthened structure, *IEEE Transactions on Industrial Electronics*, 69(1) (2021) 642–651.

[43] M. Almanza, F. Clavica, J. Chavanne, D. Moser, D. Obrist, T. Carrel, Y. Civet, Y. Perriard, Feasibility of a dielectric elastomer augmented aorta, *Advanced Science*, 8(6) (2021) 2001974.

[44] A. Walter, T. Martinez, J. Chavanne, Y. Civet, Y. Perriard, Improved electrical behavior of dielectric elastomer actuators, in: *2021 24th International Conference on Electrical Machines and Systems (ICEMS)*, IEEE, 2021, pp. 95–98.

[45] N. Liu, T. Martinez, A. Walter, Y. Civet, Y. Perriard, Control-oriented modeling and analysis of tubular dielectric elastomer actuators dedicated to cardiac assist devices, *IEEE Robotics and Automation Letters*, 7(2) (2022) 4361–4367.

[46] T. Martinez, S.E. Jahren, A. Walter, J. Chavanne, F. Clavica, L. Ferrari, P.P. Heinisch, D. Casoni, A. Haeberlin, M.M. Luedi, A novel soft cardiac assist device based on a dielectric elastomer augmented aorta: An in vivo study, *Bioengineering & Translational Medicine*, 8(2) (2023) e10396.

[47] J. Chavanne, J. Haenni, T. Martinez, D. Moser, A. Walter, M. Almanza, F. Clavica, Y. Civet, Y. Perriard, Manufacturing and tests of a tubular multilayer dielectric elastomer actuator for an impedance pump, in: *2020 23rd International Conference on Electrical Machines and Systems (ICEMS)*, IEEE, 2020, pp. 380–383.

[48] A. Benouhiba, A. Walter, S.E. Jahren, T. Martinez, F. Clavica, D. Obrist, Y. Civet, Y. Perriard, Dielectric elastomer actuator-based valveless impedance-driven pumping for meso-and macroscale applications, *Soft Robotics*, 11 (2023) 198–206.

[49] S. Wang, Z. Chen, Modeling of jellyfish-inspired robot enabled by dielectric elastomer, *International Journal of Intelligent Robotics and Applications*, 5(3) (2021) 287–299.

[50] H. Zhao, A.M. Hussain, A. Israr, D.M. Vogt, M. Duduta, D.R. Clarke, R.J. Wood, A wearable soft haptic communicator based on dielectric elastomer actuators, *Soft robotics*, 7(4) (2020) 451–461.

[51] D.-Y. Lee, S.H. Jeong, A.J. Cohen, D.M. Vogt, M. Kollosche, G. Lansberry, Y. Mengüç, A. Israr, D.R. Clarke, R.J. Wood, A wearable textile-embedded dielectric elastomer actuator haptic display, *Soft Robotics*, 9(6) (2022) 1186–1197.

[52] S. Lee, M. Moghani, A. Li, M. Duduta, A small steerable tip based on dielectric elastomer actuators, *IEEE Robotics and Automation Letters*, 8 (2023) 6531–6538.

[53] D. Niteesh, S. Malkurthi, C. Goyal, A.M. Hussain, Fabrication and characterization of a dielectric elastomer actuator based flapping wing, in: *2023 IEEE International Conference on Flexible and Printable Sensors and Systems (FLEPS)*, IEEE, 2023, pp. 1–4.

[54] B. Du, C. Tang, S. Jiang, Y. Wang, X.-J. Liu, H. Zhao, High-speed rotary motor for multidomain operations driven by resonant dielectric elastomer actuators, *Advanced Intelligent Systems*, 5(11) (2023) 2300243.

[55] J. Son, S. Lee, G.Y. Bae, G. Lee, M. Duduta, K. Cho, Skin-mountable vibrotactile stimulator based on laterally multilayered dielectric elastomer actuators, *Advanced Functional Materials*, 33(23) (2023) 2213589.

[56] C. Zhang, C. Zhang, J. Qu, X. Qian, Underwater and surface aquatic locomotion of soft biomimetic robot based on bending rolled dielectric elastomer actuators, in: *2023 IEEE/RSJ International Conference on Intelligent Robots and Systems (IROS)*, IEEE, 2023, pp. 4677–4682.

[57] F. Zhou, X. Yang, Y. Xiao, Z. Zhu, T. Li, Z. Xu, Electromechanical analysis and simplified modeling of dielectric elastomer multilayer bending actuator, *AIP Advances*, 10(5) (2020) 055003.

[58] K. Luo, Q. Tian, H. Hu, Dynamic modeling, simulation and design of smart membrane systems driven by soft actuators of multilayer dielectric elastomers, *Nonlinear Dynamics*, 102 (2020) 1463–1483.

[59] J. Mertens, O. Cabuk, J. Maas, Development of an application-oriented design tool for multilayer stack actuators based on dielectric elastomer materials, in: *ACTUATOR 2022; International Conference and Exhibition on New Actuator Systems and Applications*, VDE, 2022, pp. 1–4.

[60] M. Majidi, M. Asgari, Nonlinear electromechanical responses in multi-layered fiber-reinforced dielectric elastomer composites, *Thin-Walled Structures*, 197 (2024) 111599.

[61] X. Zhao, Z. Suo, Theory of dielectric elastomers capable of giant deformation of actuation, *Physical Review Letters*, 104(17) (2010) 178302.

[62] A. Ramaratnam, N. Jalili, Feasibility study of actuators and sensors using electroactive polymers reinforced with carbon nanotubes, in: *Smart Structures and Materials 2004: Electroactive Polymer Actuators and Devices (EAPAD)*, SPIE, 2004, pp. 349–356.

[63] V. Tomer, C. Randall, High field dielectric properties of anisotropic polymer-ceramic composites, *Journal of Applied Physics*, 104(7) (2008) 074106.

[64] B. Li, H. Chen, J. Zhou, Electromechanical stability of dielectric elastomer composites with enhanced permittivity, *Composites Part A: Applied Science and Manufacturing*, 52 (2013) 55–61.

[65] H. Stoyanov, M. Kollosche, D. McCarthy, A. Becker, S. Risse, G. Kofod, Sub-percolative composites for dielectric elastomer actuators, in: *Second International Conference on Smart Materials and Nanotechnology in Engineering*, SPIE, 2009, pp. 211–218.

[66] G. Kofod, H. Stoyanov, M. Kollosche, S. Risse, H. Ragusch, D. Rychkov, M. Dansachmüller, D. McCarthy, Nano-scale materials science for soft dielectrics: Composites for dielectric elastomer actuators, in: *2010 10th IEEE International Conference on Solid Dielectrics*, IEEE, 2010, pp. 1–4.

[67] M. Molberg, Y. Leterrier, C.J. Plummer, C. Löwe, D.M. Opris, F. Clemens, J.-A.E. Månson, Elastomer actuators: systematic improvement in properties by use of composite materials, in: *Electroactive Polymer Actuators and Devices (EAPAD) 2010*, SPIE, 2010, pp. 159–167.

[68] W. Li, C.M. Landis, Deformation and instabilities in dielectric elastomer composites, *Smart materials and structures*, 21(9) (2012) 094006.

[69] J.E.Q. Quinsaat, M. Alexandru, F.A. Nüesch, H. Hofmann, A. Borgschulte, D.M. Opris, Highly stretchable dielectric elastomer composites containing high volume fractions of silver nanoparticles, *Journal of Materials Chemistry A*, 3(28) (2015) 14675–14685.

[70] X. Sui, W. Zhou, L. Dong, Z. Wang, X. Liu, A. Zhou, J. Cai, Q. Chen, A novel fiber-reinforced silicone rubber composite with Al particles for enhanced dielectric and thermal properties, *Advances in Polymer Technology*, 37(5) (2018) 1507–1516.

[71] J.P. Szabo, J.A. Hiltz, C.G. Cameron, R.S. Underhill, J. Massey, B. White, J. Leidner, Elastomeric composites with high-dielectric constant for use in Maxwell stress actuators, in: *Smart Structures and Materials 2003: Electroactive Polymer Actuators and Devices (EAPAD)*, SPIE, 2003, pp. 180–190.

[72] N.C. Goulbourne, Cylindrical dielectric elastomer actuators reinforced with inextensible fibers, in: *Smart Structures and Materials 2006: Electroactive Polymer Actuators and Devices (EAPAD)*, SPIE, 2006, pp. 87–98.

[73] N. Goulbourne, A mathematical model for cylindrical, fiber reinforced electro-pneumatic actuators, *International Journal of Solids and Structures*, 46(5) (2009) 1043–1052.

[74] A.M. Kolesnikov, Finite deformations of a nonlinearly elastic electrosensitive tube reinforced by two fiber families, *Continuum Mechanics and Thermodynamics*, 34(5) (2022) 1237–1255.

[75] L. He, J. Lou, J. Du, Analytical solutions for inextensible fiber-reinforced dielectric elastomer torsional actuators, *Journal of Applied Mechanics*, 84(5) (2017).

[76] L. He, J. Lou, J. Du, H. Wu, Voltage-induced torsion of a fiber-reinforced tubular dielectric elastomer actuator, *Composites Science and Technology*, 140 (2017) 106–115.

[77] L. He, J. Lou, J. Du, H. Wu, Voltage-driven nonuniform axisymmetric torsion of a tubular dielectric elastomer actuator reinforced with one family of inextensible fibers, *European Journal of Mechanics-A/Solids*, 71 (2018) 386–393.

[78] L. Tian, L. Tevet-Deree, G. deBotton, K. Bhattacharya, Dielectric elastomer composites, *Journal of the Mechanics and Physics of Solids*, 60(1) (2012) 181–198.

[79] H. Yong, X. He, Y. Zhou, Electromechanical instability in anisotropic dielectric elastomers, *International Journal of Engineering Science*, 50(1) (2012) 144–150.

[80] V. Lefevre, O. Lopez-Pamies, The overall elastic dielectric properties of fiber-strengthened/weakened elastomers, *Journal of Applied Mechanics*, 82(11) (2015) 111009.

[81] R. Xiao, X. Gou, W. Chen, Suppression of electromechanical instability in fiber-reinforced dielectric elastomers, *AIP Advances*, 6(3) (2016).

[82] A.K. Sharma, M. Joglekar, A numerical framework for modeling anisotropic dielectric elastomers, *Computer Methods in Applied Mechanics and Engineering*, 344 (2019) 402–420.

[83] A. Kanan, M. Kaliske, Finite element modeling of electro-viscoelasticity in fiber reinforced electro-active polymers, *International Journal for Numerical Methods in Engineering*, 122(8) (2021) 2005–2037.
[84] A. Ahmadi, M. Asgari, Nonlinear coupled electro-mechanical behavior of a novel anisotropic fiber-reinforced dielectric elastomer, *International Journal of Non-Linear Mechanics*, 119 (2020) 103364.
[85] M. Majidi, M. Asgari, Rate-dependent electromechanical behavior of anisotropic fiber-reinforced dielectric elastomer based on a nonlinear continuum approach: modeling and implementation, *The European Physical Journal Plus*, 138(1) (2023) 1–29.
[86] K.B. Subramani, R.J. Spontak, T.K. Ghosh, Influence of fiber characteristics on directed electroactuation of anisotropic dielectric electroactive polymers with tunability, *Composites Science and Technology*, 154 (2018) 187–193.
[87] L.Z. Lyu, S. Zhu, The electromechanical behavior of dielectric elastomer actuator stiffened by fiber, *Key Engineering Materials*, 765 (2018) 12–15.
[88] E. Allahyari, M. Asgari, Fiber reinforcement characteristics of anisotropic dielectric elastomers: A constitutive modeling development, *Mechanics of Advanced Materials and Structures*, 29 (2021) 1–15.
[89] E. Allahyari, M. Asgari, Effect of fibers configuration on nonlinear vibration of anisotropic dielectric elastomer membrane, *International Journal of Applied Mechanics*, 12(10) (2020) 2050114.
[90] K. Kashyap, A.K. Sharma, M.M. Joglekar, Nonlinear dynamic analysis of aniso-visco-hyperelastic dielectric elastomer actuators, *Smart Materials and Structures*, 29(5) (2020) 055014.
[91] E. Allahyari, M. Asgari, Nonlinear dynamic analysis of anisotropic fiber-reinforced dielectric elastomers: A mathematical approach, *Journal of Intelligent Material Systems and Structures*, 32(18–19) (2021) 2300–2324.
[92] A. Alibakhshi, S. Dastjerdi, N. Fantuzzi, S. Rahmanian, Nonlinear free and forced vibrations of a fiber-reinforced dielectric elastomer-based microbeam, *International Journal of Non-Linear Mechanics*, 144 (2022) 104092.
[93] A. Moss, M. Krieg, K. Mohseni, Modeling and characterizing a fiber-reinforced dielectric elastomer tension actuator, *IEEE Robotics and Automation Letters*, 6(2) (2021) 1264–1271.
[94] C. Zeng, X. Gao, Stability of an anisotropic dielectric elastomer plate, *International Journal of Non-Linear Mechanics*, 124 (2020) 103510.
[95] A.K. Sharma, M.M. Joglekar, Effect of anisotropy on the dynamic electromechanical instability of a dielectric elastomer actuator, *Smart Materials and Structures*, 28(1) (2018) 015006.
[96] A.K. Sharma, N. Sheshkar, A. Gupta, Static and dynamic stability of dielectric elastomer fiber composites, *Materials Today: Proceedings*, 44 (2021) 2043–2047.
[97] M. Tewary, T. Roy, Electro-thermo-mechanical instability study of fiber-reinforced bimorph convex dielectric elastomer, *Soft Materials*, 21(2) (2023) 191–205.
[98] A. Baranwal, P.K. Agnihotri, Harnessing fiber induced anisotropy in design and fabrication of soft actuator with simultaneous bending and twisting actuations, *Composites Science and Technology*, 230 (2022) 109724.
[99] M. Majidi, M. Asgari, Nonlinear bending–twisting coupling in electromechanical finite deformation of fiber-reinforced tubular dielectric elastomer for soft actuators, *International Journal of Non-Linear Mechanics*, 156 (2023) 104480.
[100] T. Lu, Z. Shi, Q. Shi, T. Wang, Bioinspired bicipital muscle with fiber-constrained dielectric elastomer actuator, *Extreme Mechanics Letters*, 6 (2016) 75–81.

[101] L. Liu, C. Zhang, M. Luo, X. Chen, D. Li, H. Chen, A biologically inspired artificial muscle based on fiber-reinforced and electropneumatic dielectric elastomers, *Smart Materials and Structures*, 26(8) (2017) 085018.

[102] J. Zhang, H. Chen, D. Li, Effect of constrained fibers on electromechanical actuation of charge-controlled dielectric elastomers, *Europhysics Letters*, 120(6) (2018) 67001.

[103] S. Konstantinidi, J. Asboth, A. Walter, S. Holzer, T. Martinez, Y. Civet, Y. Perriard, An experimental approach for the design of uni-axial fiber reinforced dielectric elastomer actuators, in: *2023 26th International Conference on Electrical Machines and Systems (ICEMS)*, IEEE, 2023, pp. 3790–3794.

[104] S. Konstantinidi, T. Martinez, B. Tandon, Y. Civet, Y. Perriard, Uni-axial reinforced dielectric elastomer actuators with embedded 3D printed fibers, *Smart Materials and Structures*, 32(12) (2023) 125011.

[105] C. Li, Y. Xie, G. Li, X. Yang, Y. Jin, T. Li, Electromechanical behavior of fiber-reinforced dielectric elastomer membrane, *International Journal of Smart and Nano Materials*, 6(2) (2015) 124–134.

[106] T. Li, Y. Xie, C. Li, X. Yang, Y. Jin, J. Liu, X. Huang, Fiber-reinforced dielectric elastomer laminates with integrated function of actuating and sensing, in: *Electroactive Polymer Actuators and Devices (EAPAD) 2015*, SPIE, 2015, pp. 452–458.

[107] C. Bolzmacher, J. Biggs, M. Srinivasan, Flexible dielectric elastomer actuators for wearable human-machine interfaces, in: *Smart Structures and Materials 2006: Electroactive Polymer Actuators and Devices (EAPAD)*, SPIE, 2006, pp. 27–38.

[108] S. Shian, K. Bertoldi, D.R. Clarke, Use of aligned fibers to enhance the performance of dielectric elastomer inchworm robots, in: *Electroactive Polymer Actuators and Devices (EAPAD) 2015*, SPIE, 2015, pp. 417–425.

[109] L. Liu, J. Zhang, M. Luo, H. Chen, Z. Yang, D. Li, P. Li, A bio-inspired soft-rigid hybrid actuator made of electroactive dielectric elastomers, *Applied Materials Today*, 21 (2020) 100814.

[110] N. Sholl, A. Moss, M. Krieg, K. Mohseni, Controlling the deformation space of soft membranes using fiber reinforcement, *The International Journal of Robotics Research*, 40(1) (2021) 178–196.

[111] A. Ahmadi, M. Asgari, Novel bio-inspired variable stiffness soft actuator via fiber-reinforced dielectric elastomer, inspired by Octopus bimaculoides, *Intelligent Service Robotics*, 14(5) (2021) 691–705.

[112] M. Tewary, T. Roy, Nonlinear dynamic analysis of anisotropic bimorph dielectric elastomer actuator for soft fish robots, *Communications in Nonlinear Science and Numerical Simulation*, 127 (2023) 107585.

# Part II

## Mathematical Modeling

# 3 Hyperelasticity Frame Work for Electromechanical Characteristics

## 3.1 HYPERELASTICITY

In a general definition, hyperelasticity is the knowledge of analyzing the behavior of nonlinear elastic materials in large deformations. Contrary to classical relationships used for elastic materials in small strains, in hyperelastic relationships, stress is not obtained directly from strain.

Hyperelastic materials, such as rubber, are typically composed of long molecular chains. The molecules are cross-linked and create molecular networks. When the material is relaxed, the chains are coiled. The application of a load orders the chain and creates deformation. The network attains its original arrangement as the load is released. This explains how hyperplastic material undergo large deformation (usually up to more than 100% strain), reversibly. The deformation in hyperelastic materials usually remains elastic. However, the stress-strain curve is nonlinear, and Hooke's laws cannot be used. Therefore, stress terms are extracted from a strain energy function written based on the large deformations. Resulted equations that describe the behavior of a deformable material using field functions such as strain are known as constitutive law.

Dielectric elastomers show a strong nonlinear behavior under the application of electric voltage due to having very high strains. Therefore, one needs the principle of virtual work and potential energy function of strain W to obtain stress.

## 3.2 LARGE DEFORMATION OF HYPERELASTIC MATERIALS

The theory of continuum mechanics assumes an elastic continuum body, represented by $\mathcal{B}_0$, to be a continuous distribution of matter in space and time. The mentioned object is a collection of particles defined as $P \in \mathcal{B}_0$.

Occupied regions by the body at any given time are known as configurations. The body $\mathcal{B}_0$ occupies region $\Omega_0$ at a fixed reference time. Therefore, $\Omega_0$ depicts the reference configuration. An arbitrary particle of the body, $p$, in reference configuration, has the position vector $\boldsymbol{X}$. Figure 3.1 depicts the detail of this transformation.

DOI: 10.1201/9781003571469-5

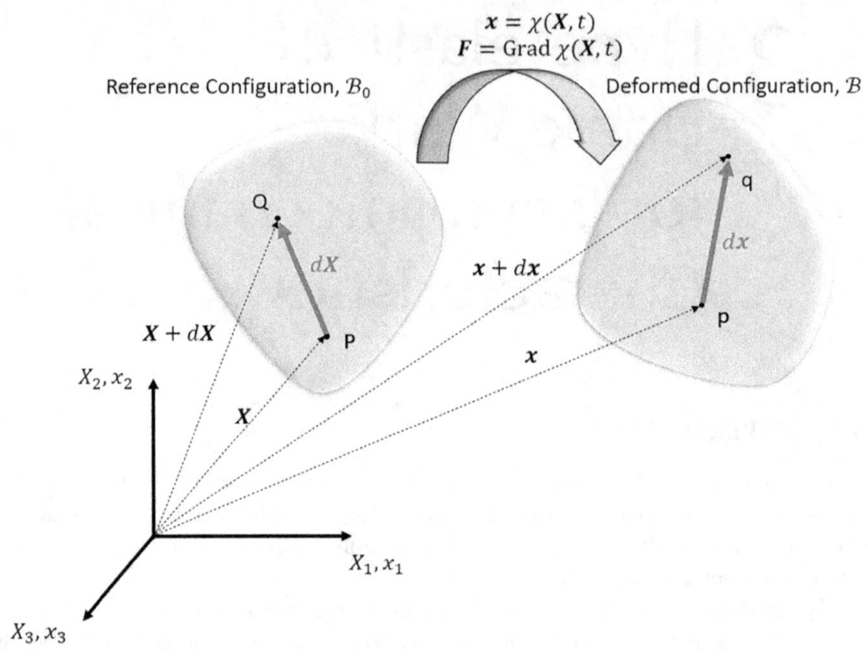


**FIGURE 3.1** Reference and current configurations of a hyperelastic material.

On the other hand, the region occupied by the body at $t = 0$ is called the initial configuration. It is commonly accepted that initial and reference configurations coincide. Then $\Omega_0$ represents the space occupied by the undeformed body at $t = 0$.

The body moves and occupies the region $\Omega$ in t ($t > 0$) in the current configuration. The motion, $\chi$, maps the particle, P in $\boldsymbol{X}$, to $\boldsymbol{x}$.

### 3.2.1 Deformation Gradient

Consider a curve $\boldsymbol{X}$ moving from refence configuration to $\boldsymbol{x}$ in current configuration. Contrary to $\boldsymbol{x}$, the material configuration of the curve, $\boldsymbol{X}$, is not a function of time. Therefore, $\boldsymbol{X}$ and $\boldsymbol{x}$ are defined as:

$$\boldsymbol{X} = X(\xi) \tag{3.1}$$

$$\boldsymbol{x} = x(\xi, t) \tag{3.2}$$

$\xi$ is the basic vector of the system.

The motion is ruled by $\chi$:

$$\boldsymbol{x} = \chi(\boldsymbol{X}, t) = \chi(X(\xi), t) \tag{3.3}$$

Tangent vectors of the curve are denoted as:

$$dx = x'(\xi,t)d\xi \tag{3.4}$$

$$dX = X'(\xi)d\xi$$

where $(\bullet)' = \partial(\bullet)/\partial\xi$. Using the chain rule, we have

$$dx = \frac{\partial\chi(X,t)}{\partial X}\frac{dX}{d\xi}d\xi = \frac{\partial\chi(X,t)}{\partial X}dX \tag{3.5}$$

Now we introduce the deformation gradient, **F**, as

$$F(X,t) = \frac{\partial\chi(X,t)}{\partial X} = \text{Grad}\,x(X,t) \tag{3.6}$$

*F* quantifies the motion in the neighborhood of a point. Using the material (reference) coordinates $X_A, A = 1,2,3$ and spatial (current) coordinates, $x_a, a = 1,2,3$, we can calculate all the nine components of **F** at any given time, *t*:

$$F = \begin{bmatrix} \frac{\partial x_1}{\partial X_1} & \frac{\partial x_1}{\partial X_2} & \frac{\partial x_1}{\partial X_3} \\ \frac{\partial x_2}{\partial X_1} & \frac{\partial x_2}{\partial X_2} & \frac{\partial x_2}{\partial X_3} \\ \frac{\partial x_3}{\partial X_1} & \frac{\partial x_3}{\partial X_2} & \frac{\partial x_3}{\partial X_3} \end{bmatrix} \tag{3.7}$$

### 3.2.2 Jacobian Determinant

As the body deforms from the reference configuration to the current configuration, its volume changes. The volume ratio, *J* that quantifies volume changes is defined as:

$$J(X,t) = \frac{dv}{dV} = \det F(X,t) \tag{3.8}$$

where $dV = dX_1dX_2dX_3$ and $dv = dx_1dx_2dx_3$ are the body volume in reference and current configuration, respectively.

*J* also represent deformation gradient determinant. We assume *F* is reversible therefore $J \neq 0$. On the other hand, the volume of a body cannot be negative. Therefore $J(X,t) > 0$.

If a motion does not change the volume, then $\mathbf{F} = \mathbf{I}$ and $J = 1$. Such a motion is an isochoric deformation.

### 3.2.3 Nanson's Formula

Due to the motion, a unit vector, normal to a surface in the reference configuration, $\boldsymbol{N}$ is mapped to a new position. However, the new vector, $\boldsymbol{n}$, would not be essentially the unit vector normal to the current surface. Nanson's formula is supposed to define the relationship between the surfaces and the normal vectors.

First, we define the volume element in the current configuration using the dot product:

$$dv = d\boldsymbol{s}.d\boldsymbol{x} \tag{3.9}$$

where, $d\boldsymbol{s} = ds\boldsymbol{n}$.

Using the volume ratio, (3.9) reforms to

$$dv = J\,dV = J\,d\boldsymbol{S}.d\boldsymbol{X} \tag{3.10}$$

where, $d\boldsymbol{S} = dS\boldsymbol{N}$.

Combining (3.9) and (3.10) results to

$$dv = d\boldsymbol{s}.d\boldsymbol{x} = J\,d\boldsymbol{S}.d\boldsymbol{X} \tag{3.11}$$

Substituting $d\mathbf{x}$ with $\mathbf{F}d\mathbf{X}$, we have

$$dv = d\boldsymbol{s}.\mathbf{F}\,d\boldsymbol{X} = J\,d\boldsymbol{S}.d\boldsymbol{X} \tag{3.12}$$

Then

$$\left(d\boldsymbol{s}.\boldsymbol{F} - J\,d\boldsymbol{S}\right)d\boldsymbol{X} = \mathbf{0} \tag{3.13}$$

Finally, the vector element of areas in reference and current configurations are related using Nanson's formula:

$$d\boldsymbol{s} = J\boldsymbol{F}^{-\mathrm{T}}d\boldsymbol{S} \tag{3.14}$$

### 3.2.4 Stress and Strain Tensors

To measure the changes of a unit vector $\boldsymbol{a}$ at $\boldsymbol{X}$ due to the deformation, we define a stretch vector, $\boldsymbol{\lambda}_a$:

$$\boldsymbol{\lambda}_a\left(\boldsymbol{X},t\right) = \boldsymbol{F}\left(\boldsymbol{X},t\right)\boldsymbol{a} \tag{3.15}$$

Stretch ratio, $\lambda$, calculated as $\lambda = |\boldsymbol{\lambda}_a|$ represent length changes of the unit vector $\boldsymbol{a}$: $\lambda > 1$ for extension, $\lambda < 1$ for compression and $\lambda = 1$ for no changes in length.

λ is used to define essential strain tensors, as well:

$$\lambda^2 = \lambda_a.\lambda_a = F(X,t)a.F(X,t)a = a.F^T Fa = a.Ca \quad (3.16)$$

where **C** is the right Cauchy-Green deformation tensor:

$$C = F^T F \quad (3.17)$$

Left Cauchy-Green deformation tensor, **B**, is defined as:

$$B = FF^T \quad (3.18)$$

On the other hand, we assume a force, *df*, is acting on the deformable body. It introduces Cauchy (true) traction stress, **t** and Piola-Kirchhoff (nominal) traction stress, **T** per unit surface area in reference and current configurations:

$$df = tds = Tds \quad (3.19)$$

Corresponding second-order tensor fields are:

$$t(x;t;n) = \sigma(x;t)n \quad (3.20)$$

$$T(X;t;N) = P(X;t)N \quad (3.21)$$

where $\sigma$ is a spatially symmetric tensor field called true or Cauchy stress tensor, and $P$ is the first Piola-Kirchhoff stress tensor. The stress traction vectors and tensors in the reference and current configurations are related as follows:

$$t(x;t;n)ds = T(X;t;N)ds \quad (3.22)$$

$$\sigma(x;t)n\,ds = P(X;t)N\,dS \quad (3.23)$$

Nanson's formula transforms the true stress tensor, $\sigma$ to the first Piola-Kirchhoff stress tensor, $P$:

$$P = J\sigma F^{-T} \quad (3.24)$$

## 3.3 CONSTITUTIVE EQUATIONS FOR HYPERELASTIC MATERIALS

A material is hyperelastic if there exists an elastic potential function W, known as Helmholtz free energy function, to capture the energy variation caused by deformation in it. Therefore, to approach the constitutive equations for a hyperelastic materials introduction of W is essential. Since W is a function of strain related tensors it can be known as strain energy function. Here, we first assume W to be written based on $\boldsymbol{F}$.

For a hyperelastic material the mechanical response function for stress is formulated as follows [1]:

$$\boldsymbol{P} = \frac{\partial W(\boldsymbol{F})}{\partial \boldsymbol{F}} \tag{3.25}$$

To extract true stress tensor, we need equation (3.26):

$$\sigma = J^{-1}\boldsymbol{P}\boldsymbol{F}^{T} = \sigma^{T} \tag{3.26}$$

$$\sigma = J^{-1}\frac{\partial W(\boldsymbol{F})}{\partial \boldsymbol{F}}\boldsymbol{F}^{T} = J^{-1}\boldsymbol{F}\left(\frac{\partial W(\boldsymbol{F})}{\partial \boldsymbol{F}}\right)^{T} \tag{3.27}$$

Mentioned equations establish the leading relations to drive the constitutive laws for a hyperlastic material. They lead to material models to study the behavior of a hyperelastic material. Since the strain energy is described as a scaler-valued function, hyperelasticity is a conservative property. As the name "strain energy function" may imply, in the reference configuration, when there is no strain (**F**=**I**), the strain energy function vanishes:

$$W = W(\boldsymbol{I}) = 0 \tag{3.28}$$

This also conveys that reference configuration does not contain residual stress, and it is stress-free.

The introduction of strain increases the strain energy function. Therefore:

$$W = W(\boldsymbol{F}) \geq 0 \tag{3.29}$$

So, any suggested strain energy function is valid in ranges that guarantee this requirement.

Using the chain rule to derivate the strain energy function with respect to time leads to:

$$\dot{W} = tr\left[\left(\frac{\partial W(\boldsymbol{F})}{\partial \boldsymbol{F}}\right)^{T}\dot{\boldsymbol{F}}\right] = tr\left[\left(\frac{\partial W(\boldsymbol{C})}{\partial \boldsymbol{C}}\right)\dot{\boldsymbol{C}}\right]$$

$$= tr\left[\frac{\partial W(\boldsymbol{C})}{\partial \boldsymbol{C}}\left(\dot{\overline{\boldsymbol{F}^{T}}}\boldsymbol{F} + \boldsymbol{F}^{T}\dot{\boldsymbol{F}}\right)\right] = 2\,tr\left[\frac{\partial W(\boldsymbol{C})}{\partial \boldsymbol{C}}\boldsymbol{F}^{T}\dot{\boldsymbol{F}}\right] \tag{3.30}$$

The relation is valid for any arbitrary $\dot{\boldsymbol{F}}$. On the other hand, since $\boldsymbol{C}$ is a symmetric second-order tensor, $\frac{\partial W(\boldsymbol{C})}{\partial \boldsymbol{C}}$ is symmetric, as well. So:

$$\left(\frac{\partial W(\boldsymbol{F})}{\partial \boldsymbol{F}}\right)^T = 2\frac{\partial W(\boldsymbol{C})}{\partial \boldsymbol{C}}\boldsymbol{F}^T \tag{3.31}$$

Replacing (3.31) in (3.27), we have:

$$\boldsymbol{\sigma} = J^{-1}\boldsymbol{F}\left(\frac{\partial W(\boldsymbol{F})}{\partial \boldsymbol{F}}\right)^T = 2J^{-1}\boldsymbol{F}\frac{\partial W(\boldsymbol{C})}{\partial \boldsymbol{C}}\boldsymbol{F}^T \tag{3.32}$$

Then, for the first and second Piola-Kirchhoff stress tensors, we have:

$$\boldsymbol{P} = 2\boldsymbol{F}\frac{\partial W(\boldsymbol{C})}{\partial \boldsymbol{C}} \tag{3.33}$$

$$\boldsymbol{S} = 2\frac{\partial W(\boldsymbol{C})}{\partial \boldsymbol{C}} \tag{3.34}$$

Finally, the true stress tensor is defined based on the second Piola-Kirchhoff stress tensors

$$\boldsymbol{\sigma} = J^{-1}\boldsymbol{F}\boldsymbol{S}\boldsymbol{F}^T \tag{3.35}$$

### 3.3.1 Constitutive Equations in Term of Strain Tensor Invariants

For an isotropic material, the strain energy function may be written based on independent strain invariants of $\boldsymbol{C}$, $I_a = I_a(\boldsymbol{C}), a = 1,2,3$. Then we would have:

$$W = W\left(I_1(\boldsymbol{C}), I_2(\boldsymbol{C}), I_3(\boldsymbol{C})\right) \tag{3.36}$$

The strain invariants are [1]:

$$I_1 = I_1(\boldsymbol{C}) = tr(\boldsymbol{C}) = \lambda_1^2 + \lambda_2^2 + \lambda_3^2 \tag{3.37}$$

$$I_2 = I_2(\boldsymbol{C}) = \frac{1}{2}\left[\left(tr(\boldsymbol{C})\right)^2 - tr\left(\boldsymbol{C}^2\right)\right] = (\lambda_1\lambda_2)^2 + (\lambda_2\lambda_3)^2 + (\lambda_1\lambda_3)^2 \tag{3.38}$$

$$I_3 = I_3(\boldsymbol{C}) = det\ \boldsymbol{C} = (\lambda_1\lambda_2\lambda_3)^2 = 1 \tag{3.39}$$

where $\lambda_i, i = 1,2,3$ are the principal stretches. The assumption of the continuous derivative of the strain energy function with respect to the principal invariants makes it possible to use the chain rule another time:

$$\frac{\partial W(\boldsymbol{C})}{\partial \boldsymbol{C}} = \frac{\partial W}{\partial I_1}\frac{\partial I_1}{\partial \boldsymbol{C}} + \frac{\partial W}{\partial I_2}\frac{\partial I_2}{\partial \boldsymbol{C}} + \frac{\partial W}{\partial I_3}\frac{\partial I_3}{\partial \boldsymbol{C}} = \sum_{a=1}^{3}\frac{\partial W}{\partial I_a}\frac{\partial I_a}{\partial \boldsymbol{C}} \tag{3.40}$$

where

$$\frac{\partial I_1}{\partial \boldsymbol{C}} = \frac{\partial tr(\boldsymbol{C})}{\partial \boldsymbol{C}} = \frac{\partial(\boldsymbol{I}:\boldsymbol{C})}{\partial \boldsymbol{C}} = \boldsymbol{I} \tag{3.41}$$

$$\frac{\partial I_2}{\partial \boldsymbol{C}} = \frac{1}{2}\left(2\,tr(\boldsymbol{C})\boldsymbol{I} - \frac{\partial tr(\boldsymbol{C}^2)}{\partial \boldsymbol{C}}\right) = I_1\boldsymbol{I} - \boldsymbol{C} \tag{3.42}$$

$$\frac{\partial I_3}{\partial \boldsymbol{C}} = \frac{\partial \det \boldsymbol{C}}{\partial \boldsymbol{C}} = I_2\boldsymbol{I} - I_1\boldsymbol{C} + \boldsymbol{C}^2 = I_3\boldsymbol{C}^{-1} \tag{3.43}$$

According to (3.34), (3.40) and (3.41) to (3.43), we have

$$\boldsymbol{S} = 2\frac{\partial W(\boldsymbol{C})}{\partial \boldsymbol{C}} = 2\left[\left(\frac{\partial W}{\partial I_1} + I_1\frac{\partial W}{\partial I_2}\right) - \frac{\partial W}{\partial I_2}\boldsymbol{C} + I_3\frac{\partial W}{\partial I_3}\boldsymbol{C}^{-1}\right] \tag{3.44}$$

The second Piola-Kirchhoff stress tensor is transformed into the Cauchy stress tensor:

$$\boldsymbol{\sigma} = 2J^{-1}\left[\left(\frac{\partial W}{\partial I_1} + I_1\frac{\partial W}{\partial I_2}\right)\boldsymbol{B} - \frac{\partial W}{\partial I_2}\boldsymbol{B}^2 + I_3\frac{\partial W}{\partial I_3}\boldsymbol{I}\right] \tag{3.45}$$

Following the Cayley-Hamilton equation, $\boldsymbol{B}^3 - I_1\boldsymbol{B}^2 + I_2\boldsymbol{B} - I_3\boldsymbol{I} = 0$ [1], we have

$$\boldsymbol{B}^2 = I_1\boldsymbol{B} - I_2 I + I_3\boldsymbol{B}^{-1} \tag{3.46}$$

Replacing (3.46) in (3.45) results in:

$$\boldsymbol{\sigma} = 2J^{-1}\left[\left(I_2\frac{\partial W}{\partial I_2} + I_3\frac{\partial W}{\partial I_3}\right)\boldsymbol{I} + \frac{\partial W}{\partial I_1}\boldsymbol{B} - I_3\frac{\partial W}{\partial I_2}\boldsymbol{B}^{-1}\right] \tag{3.47}$$

### 3.3.2 Constitutive Equations in Term of Principal Stretches

In the case of invariant strain energy function, W can be written as a function of principal stretches $\lambda_a, a = 1,2,3$:

$$W = W(\boldsymbol{C}) = W(\lambda_1, \lambda_2, \lambda_3) \tag{3.48}$$

Then:

$$\sigma_a = J^{-1}\lambda_a \frac{\partial W}{\partial \lambda_a}, a = 1,2,3 \tag{3.49}$$

where the volume ratio is presented based on principal stretches

$$J = \lambda_1 \lambda_2 \lambda_3 \tag{3.50}$$

The principal first and second Piola-Kirchhoff are:

$$P_a = \frac{\partial W}{\partial \lambda_a}, S_a = \frac{1}{\lambda_a}\frac{\partial W}{\partial \lambda_a}, a = 1,2,3 \tag{3.51}$$

$$P_a = J\lambda_a^{-1}\sigma_a, S_a = J\lambda_a^{-2}\sigma_a, a = 1,2,3 \tag{3.52}$$

### 3.3.3 Common Strain Energy Functions

Numerous strain energy functions have been introduced till now. Each one supports particular features. The most appropriate strain energy function should be selected based on the material's properties and behavior. Here we review some of the commonly employed strain energy functions for the hyperelastic materials. These models usually contain couple of constants that should be determined using experimental results for each material.

#### 3.3.3.1 Ogden Model

This model is an appropriate choice to capture the behavior of incompressible materials. It is expressed as a function of principal stretches [2].

$$W = W(\lambda_1, \lambda_2, \lambda_3) = \sum_{i=1}^{n} \frac{\mu_i}{\alpha_i}\left(\lambda_1^{\alpha_i} + \lambda_2^{\alpha_i} + \lambda_3^{\alpha_i} - 3\right) \tag{3.53}$$

where $\mu_i$ and $\alpha_i$ are classical shear modulus in reference configuration and dimensionless model constants, respectively. n, rank of model, appositive integer determines the number of terms in the model.

Since the model constants do not represent physical concepts there is no restriction on their values. This feature makes Ogden model more flexible to follow real behavior of the interested material, compared to other ones.

In addition, Ogden model is observed to need no more than three pairs of constants (N=3) to capture the experimental results of simple tests including simple tension, pure shear and equiaxial tension.

#### 3.3.3.2 Neo-Hookean Model

For n=1 and $\alpha_1 = 2$, the Ogden model reduces to the Neo-Hookean model. This model is occasionally expressed based on the first strain invariant, $I_1$.

The Neo-Hookean model represents the changes of strain energy caused by stretches of polymer chains as [3]:

$$W = \frac{\mu}{2}\left(\lambda_1^2 + \lambda_2^2 + \lambda_3^2 - 3\right) = C_{10}\left(I_1 - 3\right) \tag{3.54}$$

This model benefits from the simplicity due to involving only one constant. However, this may also cause some limitations in correlation with experimental stress-deformation results. For instance, Neo-Hookean model cannot describe the behavior of rubbersin large deformations.

#### 3.3.3.3 Mooney-Rivlin Model

Setting N=2, $\alpha_1 = 2$ and $\alpha_2 = -2$ in Ogden mode, results in the Mooney-Rivlin model. It is a well known and used model in the hyperelastic material investigations. Depending on the available data and purpose, it may be written as a function of principal stretches or strain invariants [4, 5]:

$$W = \frac{\mu_1}{2}\left(\lambda_1^2 + \lambda_2^2 + \lambda_3^2 - 3\right) - \frac{\mu_2}{2}\left(\lambda_1^{-2} + \lambda_2^{-2} + \lambda_3^{-2} - 3\right) \tag{3.55}$$

For incompressible materials where, $I_3 = \lambda_1^2\lambda_2^2\lambda_3^2 = 1$, the model reforms to

$$W = C_{10}\left(I_1 - 3\right) + C_{01}\left(I_2 - 3\right) \tag{3.56}$$

Increasing the number of constants to two, strengthen the model to follow the experimental data. While the Neo-Hookean function captures the strains less than 50%, Mooney-Rivlin model improves the predictable strain to 90%.

It worth noting that contrary to Ogden model, constant parameters of Mooney-Rivlin and Neo-Hookean models represent physical concepts. Therefore, there are some restrictions on their values. For Mooney-Rivlin model it is:

$$\mu = 2\left(C_{10} + C_{01}\right) > 0 \tag{3.57}$$

#### 3.3.3.4 Yeoh Model

The Family of Ogden models are not sufficient to capture the highly nonlinear behavior of engineering elastomers. These elastomers usually contain some fillers in order to enhance physical properties. Fillers, including carbon black and silica, reinforce the material via forming physical and chemical bonds with polymer chains. Therefore, more complex models are needed to represent practical elastomers. Yeoh model, and some other models, are supposed to address this issue.

Yeoh [6] suggested a three-term model, attaining the assumption $\frac{\partial W}{\partial I_2} = 0$:

$$W = C_{10}(I_1 - 3) + C_{20}(I_1 - 3)^2 + C_{30}(I_1 - 3)^3 \tag{3.58}$$

Material constants, $C_{10}, C_{20}, C_{30}$ have to satisfy the following restriction:

$$\mu = 2C_{10} + 4C_{20}(I_1 - 3) + 6C_{30}(I_1 - 3)^2 > 0 \tag{3.59}$$

#### 3.3.3.5 Arruda and Boyce Model

Arruda and Boyce model is a statistical model, similar to Yeoh one. The model contains some constants which depend on the polymer chains orientations [7].

$$W = \mu \sum_{i=1}^{5} \frac{C_i}{\lambda_m^{2i-2}} (I_1^i - 3) \tag{3.60}$$

$\lambda_m$ denotes the stretch corresponding to the polymer chain network locking.

#### 3.3.3.6 Gent Model

Gent model provides a significant feature to model strain-stiffening. The strain energy function for Gent model is [8]:

$$W = -\frac{\mu J_{lim}}{2} \log\left(1 - \frac{\lambda_1^2 + \lambda_2^2 + \lambda_3^2 - 3}{J_{lim}}\right) \tag{3.61}$$

The model is introduced motivated to capture the chain stretch limit, $J_{lim}$. When $J_{lim} \to \infty$ it reduces to the Neo-Hookean model. Moreover, $\frac{\lambda_1^2 + \lambda_2^2 + \lambda_3^2 - 3}{J_{lim}} \to 1$, models the situation of reaching to stretch limit for the elastomer. Then, the strain energy diverges.

It also benefits from enough simplicity due to having only two material constants.

## 3.4 INCOMPRESSIBILITY

Materials able to sustain finite strain with no or negligible volume changes are known as incompressible materials. These experience isochoric motions that simplify governing equations and solutions. Since incompressibility is a valid assumption for dielectric elastomers, we review constitutive terms for incompressible materials.

Incompressibility is an ideal assumption which may force difficulties to the computational mechanics. To avoid this issue and utilize other privileges, it is a common idea to split the deformation of a material to two part: volumetric and isochoric parts. Incompressible materials do not have volumetric deformation.

To this aim, the deformation gradient is multiplicatively decomposed to volume changing $\left(J^{\frac{1}{3}}\boldsymbol{I}\right)$ and volume preserving $\overline{\boldsymbol{F}}$, parts:

$$\boldsymbol{F} = \overline{\boldsymbol{F}}\boldsymbol{F}_{vol} = \overline{\boldsymbol{F}}\left(J^{\frac{1}{3}}\boldsymbol{I}\right) \tag{3.62}$$

$$\overline{\boldsymbol{F}} = J^{-\frac{1}{3}}\boldsymbol{F} \tag{3.63}$$

Consequently, the modified deformation gradient is also defined:

$$\overline{\boldsymbol{C}} = \overline{\boldsymbol{F}}^T\overline{\boldsymbol{F}} = J^{-\frac{2}{3}}\boldsymbol{C} \tag{3.64}$$

The augmented form of the energy function to include volume changing and volume preserving parts separately is:

$$W = W\left(\overline{\boldsymbol{F}}\right) + W_{vol}\left(J\right) \tag{3.65}$$

Incompressibility is often represented with the constraint $J = 1$. Therefore, a usual form of $W_{vol}\left(J\right)$ is a kind of penalty term, $\hat{p}\left(J-1\right)$[9]. P may be an undetermined Lagrange multiplier. It can be interpreted as hydrostatic pressure and is defined considering equilibrium equation and boundary condition.

When the incompressible assumption is perfectly valid, term $\hat{p}\left(J-1\right)$ and $W_{vol}\left(J\right)$ which is supposed to model strain energy changes because of volume changes would vanish. Otherwise, equation (3.65) represents strain energy changes due to volumetric and isochoric changes separately.

Following the same procedure, the second Piola-Kirchhoff stress tensor is derived as:

$$\boldsymbol{S} = 2\frac{\partial W\left(\overline{\boldsymbol{C}}\right)}{\partial \boldsymbol{C}} + 2\frac{\partial W_{vol}\left(J\right)}{\partial \boldsymbol{C}} \tag{3.66}$$

And then the Cauchy stress tensor would be:

$$\boldsymbol{\sigma} = 2J^{-1}F\left(\frac{\partial W\left(\overline{\boldsymbol{C}}\right)}{\partial \boldsymbol{C}} + \frac{\partial W_{vol}\left(J\right)}{\partial \boldsymbol{C}}\right)\boldsymbol{F}^T \tag{3.67}$$

To derive the required equations, we need corresponding modified parameters:

$$\overline{I}_1 = I_1\left(\overline{\boldsymbol{C}}\right) = tr\left(\overline{\boldsymbol{C}}\right) = J^{-\frac{2}{3}}I_1 \tag{3.68}$$

$$\bar{I}_2 = I_2(\bar{C}) = \frac{1}{2}\left[\left(tr(\bar{C})\right)^2 - tr(\bar{C}^2)\right] = J^{-\frac{4}{3}} I_2 \tag{3.69}$$

$$\bar{I}_3 = I_3(\bar{C}) = \det(\bar{C}) = 1 \tag{3.70}$$

We also need some derivative with respect to $\boldsymbol{C}$

$$\frac{\partial J}{\partial \boldsymbol{C}} = \frac{J}{2}\boldsymbol{C}^{-1} \tag{3.71}$$

$$\frac{\partial J^{-\frac{2}{3}}}{\partial \boldsymbol{C}} = -\frac{1}{3} J^{-\frac{2}{3}} \boldsymbol{C}^{-1} \tag{3.72}$$

## 3.5 TRANSVERSELY ISOTROPIC MATERIALS

Imbedding some fiber families in a matrix phase is considered as an appropriate method to improve the mechanical properties of a matrix. Properties of the final composite is highly dependent on the direction of the fibers.

A matrix reinforced with a family of fibers has a preferred direction along the fibers. Stress field of the composite is determined by deformation gradient and that preferred direction. The material's behavior along the perpendicular direction is isotropic. So, the material is known as a transversely isotropic.

The direction of the fiber family at point $\mathbf{X} \in \Omega_0$ is defined using a unit vector $\boldsymbol{A}(\boldsymbol{X}), |\boldsymbol{A}| = 1$ in reference configuration. As the body deforms, the unit vector in current configuration would be $\boldsymbol{a}(\boldsymbol{x},t), |\boldsymbol{a}| = 1$.

The motion also changes fibers' length. Fibers' length changes are characterized by parameter fiber stretch, $\lambda_f$, ratio of the fiber length in reference configuration to current configuration. The following equation explains how the fiber position in reference and current configuration are related:

$$\lambda_f \boldsymbol{a}(\boldsymbol{x},t) = \boldsymbol{F}(\boldsymbol{X},t)\boldsymbol{A}(\boldsymbol{X}) \tag{3.73}$$

Considering $|\boldsymbol{a}| = 1$ we have

$$\lambda_f^2 = \boldsymbol{A}.\boldsymbol{F}^T\boldsymbol{F}\boldsymbol{A} = \boldsymbol{A}.\boldsymbol{C}\boldsymbol{A} \tag{3.74}$$

Equation (3.74) conveys that length changes of fibers depend on both fiber direction in the reference configuration and strain tensors.

Therefore, to capture the behavior of a hyperplastic material reinforced with a family of fibers, the strain energy function should depend on the fibers' direction, as well as the strain tensor. To this end, a second-order tensor as $\boldsymbol{A} \otimes \boldsymbol{A}$ is defined ($\otimes$ denotes the tensor product). Moreover, the strain energy function would be:

$$W = W(\boldsymbol{C}, \boldsymbol{A} \otimes \boldsymbol{A}) \tag{3.75}$$

To have the strain energy function based on strain invariants, two new pseudo-invariants of $\boldsymbol{C}$ (and $\boldsymbol{A} \otimes \boldsymbol{A}$) are defined [10]:

$$I_4(\boldsymbol{C}, \boldsymbol{A}) = \boldsymbol{A} \cdot \boldsymbol{C}\boldsymbol{A} \tag{3.76}$$

$$I_5(\boldsymbol{C}, \boldsymbol{A}) = \boldsymbol{A} \cdot \boldsymbol{C}^2\boldsymbol{A} \tag{3.77}$$

These parameters represent fibers' properties and their interaction with the hyperelastic matrix.

Then strain energy function for a transversely isotropic material composed of a fibers' family and hyperelastic matrix is expressed as:

$$W = W\left(I_1(\boldsymbol{C}), I_2(\boldsymbol{C}), I_3(\boldsymbol{C}), I_4(\boldsymbol{C}, \boldsymbol{A}), I_5(\boldsymbol{C}, \boldsymbol{A})\right) \tag{3.78}$$

Similar to the previous ones, the second Piola-Kirchhoff tensor is derived:

$$\boldsymbol{S} = 2\frac{\partial W(\boldsymbol{C}, \boldsymbol{A} \otimes \boldsymbol{A})}{\partial \boldsymbol{C}} = 2\sum_{a=1}^{5} \frac{\partial W(\boldsymbol{C}, \boldsymbol{A} \otimes \boldsymbol{A})}{\partial I_a}\frac{\partial I_a}{\partial \boldsymbol{C}} \tag{3.79}$$

Derivative of the new pseudo-invariants with respect to **C** is required:

$$\frac{\partial I_4}{\partial C} = \boldsymbol{A} \otimes \boldsymbol{A} \tag{3.80}$$

$$\frac{\partial I_5}{\partial \boldsymbol{C}} = \boldsymbol{A} \otimes \boldsymbol{C}\boldsymbol{A} + \boldsymbol{A}\boldsymbol{C} \otimes \boldsymbol{A} \tag{3.81}$$

Combining equations (3.79), (3.80), (3.81) and former introduced relations leads to the final form of the second Piola-Kirchhoff stress tensor:

$$\begin{aligned} \boldsymbol{S} = 2\Bigg[&\left(\frac{\partial W}{\partial I_1} + I_1\frac{\partial W}{\partial I_2}\right)\boldsymbol{I} - \frac{\partial W}{\partial I_2}\boldsymbol{C} + I_3\frac{\partial W}{\partial I_3}\boldsymbol{C}^{-1} + \frac{\partial W}{\partial I_4}\boldsymbol{A} \otimes \boldsymbol{A} \\ &+ \frac{\partial W}{\partial I_5}\left(\boldsymbol{A} \otimes \boldsymbol{C}\boldsymbol{A} + \boldsymbol{A}\boldsymbol{C} \otimes \boldsymbol{A}\right)\Bigg] \end{aligned} \tag{3.82}$$

It could be transformed to the Cauchy stress tensor:

$$\begin{aligned} \sigma = 2J^{-1}\Bigg[&\left(\frac{\partial W}{\partial I_1} + I_1\frac{\partial W}{\partial I_2}\right)\boldsymbol{b} - \frac{\partial W}{\partial I_2}\boldsymbol{b}^2 + I_3\frac{\partial W}{\partial I_3}\boldsymbol{I} + I_4\frac{\partial W}{\partial I_4}\boldsymbol{A} \otimes \boldsymbol{A} \\ &+ I_4\frac{\partial W}{\partial I_5}\left(\boldsymbol{A} \otimes \boldsymbol{b}\boldsymbol{A} + \boldsymbol{A}\boldsymbol{b} \otimes \boldsymbol{A}\right)\Bigg] \end{aligned} \tag{3.83}$$

## 3.6 ANISOTROPIC HYPERELASTIC MATERIAL REINFORCED WITH TWO FIBERS FAMILIES

In this section we assume the hyperelastic matrix to be reinforced with two fibers families. The fibers families unit vector in the reference configuration are represented as $\boldsymbol{A}^{(1)}$ and $\boldsymbol{A}^{(2)}$. These unit vectors would be $\boldsymbol{a}^{(1)}$ and $\boldsymbol{a}^{(2)}$ in current configuration. Details are displayed in Figure 3.2.

Equation (3.77) is generalized to cover the effect of both fibers families:

$$W = W\left(\boldsymbol{C}, \boldsymbol{A}^{(1)} \otimes \boldsymbol{A}^{(1)}, \boldsymbol{A}^{(2)} \otimes \boldsymbol{A}^{(2)}\right) \tag{3.84}$$

The pseudo-invariants unit vectors in the deformed configuration are defined as:

$$\begin{aligned} I_4\left(\boldsymbol{C}, \boldsymbol{A}^{(1)}\right) &= \boldsymbol{A}^{(1)} \cdot \boldsymbol{C}\boldsymbol{A}^{(1)}, = \boldsymbol{a}^{(1)} I_5\left(\boldsymbol{C}, \boldsymbol{A}^{(1)}\right) = \boldsymbol{A}^{(1)} \cdot \boldsymbol{C}^2 \boldsymbol{A}^{(1)} \\ I_6\left(\boldsymbol{C}, \boldsymbol{A}^{(2)}\right) &= \boldsymbol{A}^{(2)} \cdot \boldsymbol{C}\boldsymbol{A}^{(2)}, = \boldsymbol{a}^{(2)} I_7\left(\boldsymbol{C}, \boldsymbol{A}^{(2)}\right) = \boldsymbol{A}^{(2)} \cdot \boldsymbol{C}^2 \boldsymbol{A}^{(2)} \end{aligned} \tag{3.85}$$

They are similar to pseudo-invariants introduced in the previous section for the single fibers families. It is also required to have parameters to model the effect of both fibers families and their interaction.

$$I_8\left(\boldsymbol{C}, \boldsymbol{A}^{(1)}, \boldsymbol{A}^{(2)}\right) = \left(\boldsymbol{A}^{(1)} \cdot \boldsymbol{A}^{(2)}\right) \boldsymbol{A}^{(1)} \cdot \boldsymbol{C}\, \boldsymbol{A}^{(2)} \tag{3.86}$$

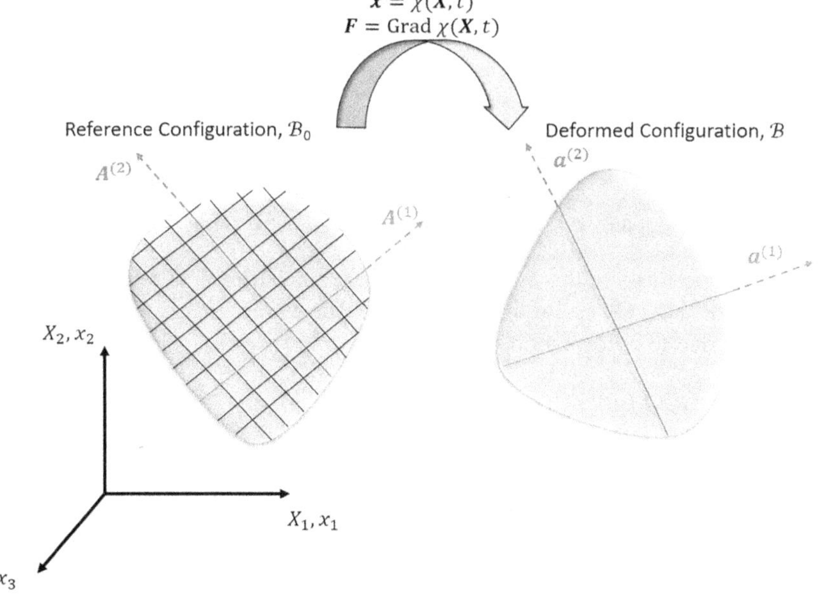


**FIGURE 3.2** Reference and current configurations of an anisotropic hyperelastic material.

$$I_9\left(C,A^{(1)},A^{(2)}\right)=\left(A^{(1)}\cdot A^{(2)}\right)^2 \tag{3.87}$$

It worth noting that $A^{(1)}\cdot A^{(2)}$ defines a geometrical constant equal to the cosine of the angle between fibers families. Therefore $I_9$ will not vary with deformation and is omitted from the following equations.

To drive the second Piola-Kirchhoff stress tensor, we need a derivative of the new pseudo-invariants with respect to $C$:

$$\frac{\partial I_4}{\partial C}=A^{(1)}\otimes A^{(1)},\ \frac{\partial I_6}{\partial C}=A^{(2)}\otimes A^{(2)} \tag{3.88}$$

$$\frac{\partial I_5}{\partial C}=A^{(1)}\otimes C\,A^{(1)}+A^{(1)}C\otimes A^{(1)} \tag{3.89}$$

$$\frac{\partial I_7}{\partial C}=A^{(2)}\otimes C\,A^{(2)}+A^{(2)}C\otimes A^{(2)}$$

$$\frac{\partial I_8}{\partial C}=\frac{1}{2}\left(A^{(1)}\cdot A^{(2)}\right)\left(A^{(1)}\otimes A^{(2)}+A^{(2)}\otimes A^{(1)}\right) \tag{3.90}$$

Finally, the expression for the second Piola-Kirchhoff stress tensor is obtained as

$$\begin{aligned}S=2\Bigg[&\left(\frac{\partial W}{\partial I_1}+I_1\frac{\partial W}{\partial I_2}\right)I-\frac{\partial W}{\partial I_2}C+I_3\frac{\partial W}{\partial I_3}C^{-1}+\frac{\partial W}{\partial I_4}A^{(1)}\otimes A^{(1)}\\&+\frac{\partial W}{\partial I_5}\left(A^{(1)}\otimes CA^{(1)}+A^{(1)}C\otimes A^{(1)}\right)+\frac{\partial W}{\partial I_6}A^{(2)}\otimes A^{(2)}\\&+\frac{\partial W}{\partial I_7}\left(A^{(2)}\otimes C\,A^{(2)}+A^{(2)}C\otimes A^{(2)}\right)\\&+\frac{1}{2}\frac{\partial W}{\partial I_8}\left(A^{(1)}\cdot A^{(2)}\right)\left(A^{(1)}\otimes A^{(2)}+A^{(2)}\otimes A^{(1)}\right)\end{aligned} \tag{3.91}$$

### 3.6.1 Anisotropic Strain Energy Function

Here we review two of the most known anisotropic strain energy functions which describe the effect of reinforcing fibers on the strain energy of a hyperelastic composite due to deformation.

#### 3.6.1.1 HGO Model

In order to support the stiffening effect at high pressure, the strain energy function for a reinforced hyperelastic material is suggested to include exponential terms. HGO model, suggested by Holzapfel et.al. [11] is:

$$W_{aniso}=\frac{k_1}{2k_2}\sum_{i=4,6}\left\{\exp\left[k_2\left(\bar{I}_i-1\right)^2\right]-1\right\} \tag{3.92}$$

where $k_1 > 0$ and $k_2$ are a stress like and a dimensionless parameter, respectively. Both are model constants and are determined using experimental test data for particular materials.

Equation (3.92) express the HGO model for a hyperelastic matrix reinforced with two fibers families. In the most general form, the model would be:

$$W_{aniso} = \frac{k_1}{2k_2} \sum_{\alpha=1}^{N} \left\{ \exp\left[ k_2 \langle \bar{E}_\alpha \rangle^2 \right] - 1 \right\} \tag{3.93}$$

$$\bar{E}_\alpha = \hat{\kappa}\left(\bar{I}_1 - 3\right) + \left(1 - 3\kappa\right)\left(\bar{I}_{4(\alpha\alpha)} - 1\right) \tag{3.94}$$

Here, $\hat{\kappa}$, N, and $\bar{I}_{4(\alpha\alpha)}$ are a temperature-dependent parameter, number of fibers families, and a pseudo-invariant, respectively.

#### 3.6.1.2 Feng Model

Another strain energy model for anisotropic hyperelastic material is suggested by Feng et. al. [12]:

$$W_{aniso} = \frac{\mu}{2}\left[ \xi_1 \left(I_4 - 1\right)^2 - \xi_2 \left(I_5 - I_4^2\right) \right] \tag{3.95}$$

Here, $\xi$ is a dimensionless parameter representing the anisotropy degree. $I_4$ conveys the strain energy changes caused by fibers' deformation. Furthermore, $\left(I_5 - I_4^2\right)$ is a function of shear strain due to the interaction between fibers and matrix.

## 3.7 DIELECTRIC EFFECT CONSTITUTIVE LAW

In order to review the constitutive equations on dielectric field, let's consider vectors $\mathbb{e}$ and $\mathbb{d}$. They are Eulerian form of electric field and electric displacement in current configuration. Assuming a polarized material medium, the polarization density vector, $\mathbb{p}$, is also defined. These vectors are related according to

$$\mathbb{p} = \mathbb{d} - \varepsilon \mathbb{e} \tag{3.96}$$

where $\varepsilon$ denotes dielectric permittivity. For vacuum or non-polarizable material, the relation transforms to:

$$\mathbb{d} = \varepsilon \mathbb{e} \tag{3.97}$$

Eulerian form of fundamental differential equations governing the vectors in a dielectric material are expressed as [13]:

$$curl\ \mathbb{e} = 0 \tag{3.98}$$

$$div\ \mathbb{d} = 0 \tag{3.99}$$

To develop the electromechanical model, expressions of the parameters in Lagrangian forms are also required. Then, $\mathbb{E}$, $\mathbb{D}$, and $\mathbb{P}$, electric field, displacement, and polarization in reference configuration are obtained as:

$$\mathbb{E} = \boldsymbol{F}^T \mathbb{e} \tag{3.100}$$

$$\mathbb{D} = J\boldsymbol{F}^{-1}\mathbb{d} \tag{3.101}$$

$$\mathbb{P} = J\boldsymbol{F}^{-1}\mathbb{p} \tag{3.102}$$

Similar differential equations are valid in Lagrangian form, as well:

$$curl\,\mathbb{E} = 0 \tag{3.103}$$

$$div\,\mathbb{D} = 0 \tag{3.104}$$

The strain energy function to model the effect of electric excitation based on the introduced parameters is [14]:

$$W^{dielectric} = -\frac{1}{2}\varepsilon J C^{-1} : (\mathbb{E} \otimes \mathbb{E}) \tag{3.105}$$

The function is used to drive the second Piola-Kirchhoff stress tensor

$$\begin{aligned} S_{ij}^{dielectric} &= 2\frac{\partial W^{dielectric}}{\partial C_{ij}} = -\varepsilon \mathbb{E}_p \mathbb{E}_q \frac{\partial \left(JC_{pq}^{-1}\right)}{\partial C_{ij}} = -\mathbb{E}_p \mathbb{E}_q \varepsilon \left( \frac{\partial J}{\partial C_{ij}} C_{pq}^{-1} + \frac{\partial C_{pq}^{-1}}{\partial C_{ij}} J \right) \\ &= -\frac{1}{2}\varepsilon J \mathbb{E}_p \mathbb{E}_q \left( C_{ij}^{-1} C_{pq}^{-1} - \frac{1}{2}\left( C_{pj}^{-1} C_{qi}^{-1} + C_{pi}^{-1} C_{qj}^{-1} \right) \right) \end{aligned} \tag{3.106}$$

Pushing **S** forward results in Cauchy stress tensor

$$\sigma_{mn}^{dielectric} = \varepsilon \left( \mathbb{e}_m \mathbb{e}_n - \frac{1}{2} \mathbb{e}_l \mathbb{e}_l \delta_{mn} \right) \tag{3.107}$$

## 3.8 ELASTICITY TENSOR

A commonly used method to investigate nonlinear equations, including those in hyperelasticity, is iterative techniques. The method solves linearized constitutive equations.

To linearize the nonlinear second Piola-Kirchhoff stress tensor, its derivative with respect to **C** is derived:

$$\mathbb{C} = 2\frac{\partial \boldsymbol{S}(\boldsymbol{C})}{\partial \boldsymbol{C}} \tag{3.108}$$

According to (3.108), $\mathbb{C}$, elasticity tensor denotes the changes of stress due to strain changes. (3.108) also shows the rank of $\mathbb{C}$, to be of four.

The equation also reveals minor symmetries for $\mathbb{C}$:

$$\mathbb{C}_{ijkl} = \mathbb{C}_{jikl} = \mathbb{C}_{ijkl} \tag{3.109}$$

Since previous equations have denoted the relation between $\boldsymbol{S}$ and W in hyperelastic materials, the equation (3.108) is reformed as

$$\mathbb{C} = 4\frac{\partial^2 W(\boldsymbol{C})}{\partial \boldsymbol{C}^2} \tag{3.110}$$

Therefore, $\mathbb{C}$ of a hyperelastic material possesses major symmetries as well

$$\mathbb{C}_{ijkl} = \mathbb{C}_{klij} \tag{3.111}$$

On the other word, existence of W for a hyperelastic material guarantees major symmetry of $\mathbb{C}$ and vice versa.

The following equation expresses the transformation needed to drive the elasticity tensor in the current configuration.

$$\mathbb{c}_{mnrs} = \frac{1}{J} F_{mi} F_{nj} F_{rk} F_{sl} \mathbb{C}_{ijkl} \tag{3.112}$$

## 3.9 AN ANISOTROPIC DIELECTRIC ELASTOMER

According to the assumptions taken for the material, a dielectric elastomer actuator, which is reinforced with two families of fibers, is an anisotropic composite. The matrix displays the characteristics of a hyperelastic material. Therefore, the comprehensive strain energy function should cover all the mentioned features. A possible model for such a material can be

$$W = \bar{W} + W^{vol} + W^{dielectric} + W^{aniso} \tag{3.113}$$

where $\bar{W}$ is the strain energy function to model isochoric deformation. $W^{vol}$ models the strain energy variation due to volumetric deformation. To capture the electromechanical behavior, the comprehensive model includes $W^{dielectric}$ . These three terms cover different characteristics of the hyperelastic matrix. Finally, $W^{aniso}$ captures the reinforcing fibers' behavior.

The next step is to decide on the model for each strain energy function term. We proceed the steps and assume Mooney-Rivlin model, equation (3.105) and HGO model for $\bar{W}$, $W^{dielectric}$ and $W^{aniso}$, respectively. We also take $\frac{\kappa}{2}(J-1)^2$ for

$W^{vol}$, where $\kappa$ is the bulk modulus of the matrix. The final version of the comprehensive model would be:

$$W = C_{10}\left(\bar{I}_1 - 3\right) + C_{01}\left(\bar{I}_2 - 3\right) + \frac{\kappa}{2}\left(J-1\right)^2 - \frac{1}{2}\varepsilon J C^{-1} : \left(\mathbb{E}\otimes\mathbb{E}\right) + \frac{k_1}{2k_2}\sum_{i=4,6}\left\{\exp\left[k_2\left(\bar{I}_i - 1\right)^2\right] - 1\right\} \tag{3.114}$$

The second Piola-Kirchhoff stress tensor will be derived

$$\begin{aligned} S_{ij} = 2\Bigg( & C_{10}J^{-\frac{2}{3}}\left(\delta_{ij} - \frac{1}{3}I_1 C_{ij}^{-1}\right) + C_{01}J^{-\frac{4}{3}}\left(I_1\delta_{ij} - C_{ij} - \frac{2}{3}I_2 C_{ij}^{-1}\right) + \frac{\kappa}{2}\left(J-1\right)JC_{ij}^{-1}\Bigg) \\ & - \frac{1}{2}\varepsilon J \mathbb{E}_p \mathbb{E}_q\left(C_{ij}^{-1}C_{pq}^{-1} - \frac{1}{2}\left(C_{pj}^{-1}C_{qi}^{-1} + C_{pi}^{-1}C_{qj}^{-1}\right)\right) \\ & + 2k_1^{(1)}\left(I_4 - 1\right)\left(\exp\left(k_2^{(1)}\left(I_4-1\right)^2\right)\right)A_i^{(1)}A_j^{(1)} \\ & + 2k_1^{(2)}\left(I_6 - 1\right)\left(\exp\left(k_2^{(2)}\left(I_6-1\right)^2\right)\right)A_i^{(2)}A_j^{(2)} \end{aligned} \tag{3.115}$$

The procedure is followed by the extraction of the Cauchy stress tensor

$$\begin{aligned} \sigma_{mn} = \frac{2}{J}\Bigg( & C_{10}\left(\bar{B}_{mn} - \frac{1}{3}\bar{I}_1\delta_{mn}\right) + C_{01}\left(\bar{I}_1\bar{B}_{mn} - \bar{B}_{mk}\bar{B}_{nk} - \frac{2}{3}\bar{I}_2\delta_{mn}\right) + \frac{\kappa}{2}\left(J-1\right)J\delta_{mn}\Bigg) \\ & + \varepsilon\left(\mathrm{e}_m\mathrm{e}_n - \frac{1}{2}\mathrm{e}_l\mathrm{e}_l\delta_{mn}\right) + \frac{2}{J}\Bigg(k_1^{(1)}\left(I_4-1\right)\left(\exp\left(k_2^{(1)}\left(I_4-1\right)^2\right)\right)a_m^{(1)}a_n^{(1)} \\ & + k_1^{(2)}\left(I_6-1\right)\left(\exp\left(k_2^{(2)}\left(I_6-1\right)^2\right)\right)a_m^{(2)}a_n^{(2)}\Bigg) \end{aligned} \tag{3.116}$$

The elasticity tensor for such a material would be:

$$\begin{aligned} \mathbb{c}_{mnrs} = & \left(\frac{4}{3}\left(C_{10}\bar{I}_1 + 4C_{01}\bar{I}_2\right) + \kappa\left(2J-1\right)\right)\delta_{mn}\delta_{rs} \\ & + \left(2\bar{I}_1C_{10} + 4\bar{I}_2C_{01} + \kappa\left(1-J\right)\right)\left(\delta_{mr}\delta_{ns} + \delta_{nr}\delta_{ms}\right) \\ & - 4\left(C_{10} + 2C_{01}\bar{I}_1\right)\left(\bar{B}_{mn}\delta_{rs} + \bar{B}_{rs}\delta_{mn}\right) \\ & + 4C_{01}\left(2\left(\bar{B}_{rk}\bar{B}_{sk}\delta_{mn} + \bar{B}_{mk}\bar{B}_{nk}\delta_{rs}\right) - 2\bar{B}_{mn}\bar{B}_{rs} + \left(\bar{B}_{mr}\bar{B}_{ns} + \bar{B}_{nr}\bar{B}_{ms}\right)\right) \\ & - \frac{1}{2}\varepsilon\left(\left(\delta_{mn}\delta_{rs} - \left(\delta_{mr}\delta_{ns} + \delta_{ms}\delta_{nr}\right)\right)\mathrm{e}_l\mathrm{e}_l\right) \\ & + \varepsilon\left(\mathrm{e}_n\mathrm{e}_m\delta_{rs} + \mathrm{e}_r\mathrm{e}_s\delta_{mn} - \mathrm{e}_n\mathrm{e}_s\delta_{mr} - \mathrm{e}_m\mathrm{e}_r\delta_{ns} - \mathrm{e}_m\mathrm{e}_s\delta_{nk} - \mathrm{e}_n\mathrm{e}_r\delta_{ms}\right) \\ & + \frac{4}{J}\Bigg(k_1^{(1)}\left(1 + 2k_2^{(1)}\left(I_4-1\right)^2\right)\exp\left(k_2^{(1)}\left(I_4-1\right)^2\right)a_m^{(1)}a_n^{(1)}a_r^{(1)}a_s^{(1)} \\ & + k_1^{(2)}\left(1 + 2k_2^{(2)}\left(I_6-1\right)^2\right)\exp\left(k_2^{(2)}\left(I_6-1\right)^2\right)a_m^{(2)}a_n^{(2)}a_r^{(2)}a_s^{(2)}\Bigg) \end{aligned} \tag{3.117}$$

## 3.10 CONCLUSION

Dielectric elastomer actuators mainly manifest the known characteristics of hyperelastic material. Therefore, it is reasonably appropriate to use hyperelastic constitutive laws to study the large deformation of DEAs. Using such a framework along with nonlinear continuum mechanics approach provides an effective tool to capture the nonlinear coupled behavior of dielectric elastomer materials. This chapter presented the fundamental steps to develop a comprehensive constitutive law that considers hyperelasticity, incompressibility, and dielectric effect at the same time, from the selection of free Helmholtz energy function to explicit expressions for Piola-Kirechhoff and Cauchy stress tensors and then elasticity tensor. The following chapter will add the effect of time dependency to the model.

## REFERENCES

[1] G.A. Holzapfel, *Nonlinear solid mechanics: a continuum approach for engineering science*, Kluwer Academic Publishers Dordrecht, 2002.

[2] R. Ogden, Large deformation isotropic elasticity—on the correlation of theory and experiment for incompressible rubberlike solids, *Rubber Chemistry and Technology*, 46(2) (1973) 398–416.

[3] L. Treloar, The elasticity of a network of long-chain molecules—II, *Transactions of the Faraday Society*, 39 (1943) 241–246.

[4] M. Mooney, A theory of large elastic deformation, *Journal of Applied Physics*, 11(9) (1940) 582–592.

[5] R.S. Rivlin, Large elastic deformations of isotropic materials IV. Further developments of the general theory, Philosophical transactions of the royal society of London. *Series A, Mathematical and Physical Sciences*, 241(835) (1948) 379–397.

[6] O.H. Yeoh, Some forms of the strain energy function for rubber, *Rubber Chemistry and Technology*, 66(5) (1993) 754–771.

[7] E.M. Arruda, M.C. Boyce, A three-dimensional constitutive model for the large stretch behavior of rubber elastic materials, *Journal of the Mechanics and Physics of Solids*, 41(2) (1993) 389–412.

[8] A.N. Gent, A new constitutive relation for rubber, *Rubber Chemistry and Technology*, 69(1) (1996) 59–61.

[9] J. Simo, R. Taylor, Penalty function formulations for incompressible nonlinear elastostatics, *Computer Methods in Applied Mechanics and Engineering*, 35(1) (1982) 107–118.

[10] A.J.M. Spencer, Constitutive theory for strongly anisotropic solids, in: *Continuum theory of the mechanics of fibre-reinforced composites*, Springer, 1984, pp. 1–32.

[11] G.A. Holzapfel, T.C. Gasser, R.W. Ogden, A new constitutive framework for arterial wall mechanics and a comparative study of material models, *Journal of Elasticity and the Physical Science of Solids*, 61(1) (2000) 1–48.

[12] Y. Feng, R.J. Okamoto, R. Namani, G.M. Genin, P.V. Bayly, Measurements of mechanical anisotropy in brain tissue and implications for transversely isotropic material models of white matter, *Journal of the Mechanical Behavior of Biomedical Materials*, 23 (2013) 117–132.

[13] A. Dorfmann, R.W. Ogden, Nonlinear electroelasticity, *Acta Mechanica*, 174(3) (2005) 167–183.

[14] A. Büschel, S. Klinkel, W. Wagner, Dielectric elastomers–numerical modeling of nonlinear visco-electroelasticity, *International Journal for Numerical Methods in Engineering*, 93(8) (2013) 834–856.

# 4 Modeling of Visco-Hyperelasticity Effects

## 4.1 TIME DEPENDENCY

Due to the three-dimensional networks of long molecular chain in elastomers, elastomers manifest significant features of large deformation, nonlinearity and time dependent behavior [1]. Hyperelasticity and Helmholtz free energy function addresses the two farmer features, adequately. The last one, time-dependency, should be modeled to capture dielectric elastomer's behavior, as well.

To study the rate-dependent behavior, we need models compatible with short-term analysis. Here we present potential energy function. This method proposes tow terms for energy and consequently two expressions for stress tensors. The elastic stress is defined using the hyperelastic strain energy functions and equation introduced in the previous chapter. To define the rate-dependent stress, a dissipation potential function is considered. Derivation of the function with respect to a time-dependent parameter leads to the expression for rate dependent stress.

On the other hand, a common approach to cover long-term time dependency and related phenomena, such as stress relaxation, is to simulate the behavior of elastomers using rheological models. They are basically made of spring and dashpot elements in various configurations. The analogy between the rheological models and elastomers' behavior is used to derive the constitutive laws. Here, we use a generalized Maxwell model to express the linear response of the elastomer.

## 4.2 RATE-DEPENDENT BEHAVIOR

The stress-strain curves of dielectric elastomer materials show strain-rate dependency. Pioletti et al. proposed incorporating this effect using a viscoelastic constitutive law [2]. The constitutive law should describe the nonlinear behavior of the material and how it is affected by the large deformation. Proposed by Pioletti et al., strain rate dependency is modeled using an additional explicit variable. A common variable could be the first time derivative of right Cauchy-Green strain tensor, $\dot{\boldsymbol{C}}$. Then a viscose potential, Wv, as a function of $C$ and $\dot{C}$ is assumed in addition to elastic potential, We($C$). Consequently, the Helmholtz free energy function would be W = We($C$)+ Wv $(\boldsymbol{C}, \dot{\boldsymbol{C}})$. This approach decouples the energy incorporation of various features. The Helmhotz free energy function could be determined as a function of strain invariants $(I_i)$ and strain rate invariants $(J_i)$.

DOI: 10.1201/9781003571469-6

This section presence an example constitutive law that considers strain rate dependency through strain rate invariants.

Equation (3.6) denotes the definition of deformation gradient for a motion to map the body from the reference configuration to the current one. To capture the rate-dependent behavior, we need the derivative of $\boldsymbol{F}$ with respect to time.

Initially we introduce the spatial velocity field, $\boldsymbol{v}(\boldsymbol{x},t)$, and it's derivative with respect to spatial coordinate, $\boldsymbol{l}(x,t)$ in the current configuration:

$$\boldsymbol{v}(\boldsymbol{x},t)=\frac{\partial \boldsymbol{x}}{\partial t}=\frac{\partial \chi(\boldsymbol{X},t)}{\partial t}=\dot{\chi}(\boldsymbol{X},t) \tag{4.1}$$

$$\boldsymbol{l}(\boldsymbol{x},t)=\frac{\partial \boldsymbol{v}(\boldsymbol{x},t)}{\partial \boldsymbol{x}}=\frac{\partial \dot{\chi}(\boldsymbol{X},t)}{\partial \boldsymbol{X}}\frac{\partial \boldsymbol{X}}{\partial \boldsymbol{x}}=\frac{\partial}{\partial t}\left(\frac{\partial \chi(\boldsymbol{X},t)}{\partial \boldsymbol{X}}\right)\boldsymbol{F}^{-1}=\dot{\boldsymbol{F}}\boldsymbol{F}^{-1} \tag{4.2}$$

$\boldsymbol{l}(x,t)$ presents the relative velocity between two particles of the deformed by divided by their relative position vector.

Therefore, the time derivative of the deformation gradient, $\dot{\boldsymbol{F}}$, can be described using $\boldsymbol{l}(x,t)$. We have:

$$\dot{\boldsymbol{F}}=\boldsymbol{l}\boldsymbol{F} \tag{4.3}$$

To include the effect of time dependency, $\dot{\boldsymbol{C}}$ is also required:

$$\dot{\boldsymbol{C}}=\dot{\boldsymbol{F}}^{T}\boldsymbol{F}+\boldsymbol{F}^{T}\dot{\boldsymbol{F}}=\boldsymbol{F}^{T}\boldsymbol{l}^{T}\boldsymbol{F}+\boldsymbol{F}^{T}\boldsymbol{l}\boldsymbol{F}=\boldsymbol{F}^{T}\left(\boldsymbol{l}^{T}+\boldsymbol{l}\right)\boldsymbol{F} \tag{4.4}$$

To simplify the following equations, $\boldsymbol{D}$, the symmetric part of $\boldsymbol{l}$ is expressed

$$\boldsymbol{D}=\frac{1}{2}\left(\boldsymbol{l}+\boldsymbol{l}^{T}\right) \tag{4.5}$$

$\boldsymbol{D}$ states the rate of deformation. So, the Equation (4.4) is updated to

$$\dot{\boldsymbol{C}}=2\boldsymbol{F}^{T}\boldsymbol{D}\boldsymbol{F} \tag{4.6}$$

Similar to Equations (3.37)–(3.39), some invariants for $\dot{\boldsymbol{C}}$ are obtained:

$$J_1=\boldsymbol{I}:\dot{\boldsymbol{C}} \tag{4.7}$$

$$J_2=\frac{1}{2}\left(\boldsymbol{I}:\dot{\boldsymbol{C}}^{2}\right) \tag{4.8}$$

$$J_3=\boldsymbol{det}\left(\dot{\boldsymbol{C}}\right) \tag{4.9}$$

$$J_4=\boldsymbol{I}:\left(\boldsymbol{C}.\dot{\boldsymbol{C}}\right) \tag{4.10}$$

These invariants model strain rate effects. Additional invariant, $J_4$ represents the interaction between stress and dissipative effects of viscosity.

Time derivative of parameters, $\dot{\boldsymbol{F}}$ and $\dot{\boldsymbol{C}}$ and rate-dependent invariants, $J_i, i = 1,\ldots,4$ provide the opportunity to define the strain energy function in such a way as to describe the dissipative nature of rate-dependency. The strain energy function for rate-dependent behavior indicates dissipation potential and should satisfy some requirements: (1) It should be convex with respect to $\dot{\boldsymbol{C}}$ [2] and (2) It should vanish when $\dot{\boldsymbol{C}} = 0$ [3].

Therefore, a function describing the short-time effects of nonlinear strain-rate sensitivity, based on $\boldsymbol{C}$ and $\dot{\boldsymbol{C}}$, is required. Since the deformation does not depend on the reference frame, the free energy function is invariant for any deformation. Therefore, it could be expressed in terms of the principal invariants of $\boldsymbol{C}$ and $\dot{\boldsymbol{C}}$, $\bar{I}_i$ and $\bar{J}_i$. Here, we propose the model expressed in Equation (4.11) [3]:

$$W^{rate-dep} = \frac{1}{4}\left(\bar{I}_1 - 3\right)\left(\eta_1 \bar{J}_2 + \eta_2 \bar{J}_4^2\right) \tag{4.11}$$

where $\eta_1$ and $\eta_2$ are model constants (Pa.S) which are determined using experimental test data. $\bar{I}_1$ and $\bar{J}_i, i = 2,4$ model strain and strain-rate effects, respectively. In the reference configuration, where $\bar{I}_1 = 3$, and at zero strain rate, the rate-dependent strain energy function vanishes, ensuring no energy dissipation. It worth mentioning that the rate-dependent strain energy function is defined for volume preserving deformation, as well.

Following a procedure similar to one studied in Chapter 3, the second Piola-Kirchhoff stress tensor is defined using derivation. However, here we derive the derivative of $W^{rate-dep}$ with respect to $\dot{\boldsymbol{C}}$:

$$\boldsymbol{S}^{rate-dep} = 2\frac{\partial W^{rate-dep}}{\partial \dot{\boldsymbol{C}}} \tag{4.12}$$

Combining (4.11) and (4.12) leads to

$$S_{ij}^{rate-dep} = 2\frac{\partial W^{rate-dep}}{\partial \dot{C}_{ij}} = \frac{1}{2}\left(\bar{I}_1 - 3\right)\left(\eta_1 \frac{\partial \bar{J}_2}{\partial \dot{C}_{ij}} + 2\eta_2 \bar{J}_4 \frac{\partial \bar{J}_4}{\partial \dot{C}_{ij}}\right) \tag{4.13}$$

The derivation procedure includes the definition of some isochoric parameters and their derivative with respect to $\dot{\boldsymbol{C}}$ [1]:

$$\dot{\bar{\boldsymbol{C}}} = \dot{\bar{\boldsymbol{F}}}^T \bar{\boldsymbol{F}} + \bar{\boldsymbol{F}}^T \dot{\bar{\boldsymbol{F}}} = 2\bar{\boldsymbol{F}}^T \boldsymbol{d}\bar{\boldsymbol{F}} = J^{-\frac{2}{3}}\dot{\boldsymbol{C}} - \frac{2}{3}J^{-\frac{5}{3}}\dot{J}\boldsymbol{C} \tag{4.14}$$

$$\dot{J} = J\,\mathrm{tr}\left(\boldsymbol{d}\right) \tag{4.15}$$

$$\dot{J} = \frac{1}{2}J\boldsymbol{I} : \left(\boldsymbol{F}^{-T}\dot{\boldsymbol{C}}\boldsymbol{F}^{-1}\right) \tag{4.16}$$

$$\frac{\partial J}{\partial \boldsymbol{F}} = J\boldsymbol{F}^{-T} \tag{4.17}$$

$$\frac{\partial \dot{\bar{\boldsymbol{C}}}}{\partial \dot{\boldsymbol{C}}} = \frac{\partial\left(J^{-\frac{2}{3}}\dot{\boldsymbol{C}} - \frac{2}{3}J^{-\frac{5}{3}}\dot{J}\boldsymbol{C}\right)}{\partial \dot{\boldsymbol{C}}} = J^{-\frac{2}{3}}\frac{\partial \dot{\boldsymbol{C}}}{\partial \dot{\boldsymbol{C}}} - \frac{2}{3}J^{-\frac{5}{3}}\frac{\partial \dot{J}}{\partial \dot{\boldsymbol{C}}}\boldsymbol{C}$$

$$= J^{-\frac{2}{3}}\frac{\partial \dot{\boldsymbol{C}}}{\partial \dot{\boldsymbol{C}}} - \frac{2}{3}J^{-\frac{5}{3}}\frac{\partial \dot{J}}{\partial \dot{\boldsymbol{C}}}\boldsymbol{C} \tag{4.18}$$

$$\frac{\partial J_2}{\partial \dot{C}_{ij}} = \dot{C}_{ij} \tag{4.19}$$

$$\frac{\partial J_4}{\partial \dot{C}_{ij}} = C_{ij} \tag{4.20}$$

$$\frac{\partial \bar{J}_2}{\partial \dot{\boldsymbol{C}}} = \frac{\partial\left(\frac{1}{2}\left(\boldsymbol{I} : \dot{\bar{\boldsymbol{C}}}^2\right)\right)}{\partial \dot{\boldsymbol{C}}} = \frac{1}{2}\frac{\partial\left(\boldsymbol{I} : \left(J^{-\frac{2}{3}}\dot{\boldsymbol{C}} - \frac{2}{3}J^{-\frac{5}{3}}\dot{J}\boldsymbol{C}\right)^2\right)}{\partial \dot{\boldsymbol{C}}} = \frac{\partial \dot{\bar{\boldsymbol{C}}}}{\partial \dot{\boldsymbol{C}}} : \dot{\bar{\boldsymbol{C}}} \tag{4.21}$$

$$\frac{\partial \bar{J}_4}{\partial \dot{C}_{ij}} = \left(J^{-\frac{4}{3}}C_{ij} - \frac{1}{3}C_{ij}^{-1}\left(\bar{I}_1^2 - 2\bar{I}_2\right)\right) = J^{-\frac{4}{3}}C_{ij} - \frac{1}{3}C_{ij}^{-1}\left(\bar{I}_1^2 - 2\bar{I}_2\right) \tag{4.22}$$

where $\boldsymbol{d}$ is the corresponding term of $\boldsymbol{D}$ in the current configuration.

Therefore, the final form of the second Piola-Kirchhoff stress tensor is

$$S_{ij}^{rate-dep} = \frac{1}{2}\left(\bar{I}_1 - 3\right)\left(\eta_1\left(J^{-\frac{2}{3}}\dot{C}_{ij} - \frac{1}{3}C_{ij}^{-1}J_4\right)\right.$$
$$\left.+ 2\eta_2\bar{J}_4J^{-\frac{4}{3}}\left(J^{-\frac{4}{3}}C_{ij} - \frac{1}{3}C_{ij}^{-1}\left(\bar{I}_1^2 - 2\bar{I}_2\right)\right)\right) = \frac{1}{2}\left(\bar{I}_1 - 3\right)$$
$$\left(\eta_1J^{-\frac{2}{3}}\dot{\bar{C}}_{ij} + 2\eta_2\bar{J}_4J^{-\frac{4}{3}}C_{ij} - \frac{1}{3}\bar{J}_4\left(\eta_1 + 2\eta_2\left(\bar{I}_1^2 - 2\bar{I}_2\right)\right)C_{ij}^{-1}\right) \tag{4.23}$$

It is pushed forward to derive the Cauchy stress tensor:

$$\sigma_{mn}^{rate-dep} = \frac{1}{J}F_{mi}F_{nj}S_{ij}^{rate-dep} = \frac{1}{2J}\left(\bar{I}_1 - 3\right)\left(\eta_1J^{-\frac{2}{3}}\dot{\bar{C}}_{ij}F_{mi}F_{nj} + 2\eta_2\bar{J}_4J^{-\frac{4}{3}}C_{ij}F_{mi}F_{nj}\right.$$
$$\left.-\frac{1}{3}\bar{J}_4\left(\eta_1 + 2\eta_2\left(\bar{I}_1^2 - 2\bar{I}_2\right)\right)C_{ij}^{-1}F_{mi}F_{nj}\right)$$
$$= \frac{1}{2J}\left(\bar{I}_1 - 3\right)\left(2\eta_1\bar{B}_{mi}D_{ij}\bar{B}_{jn} + 2\left(\eta_2\bar{J}_4 - \frac{1}{3}\eta_1tr(d)\right)\bar{B}_{mk}\bar{B}_{nk}\right.$$
$$\left.-\frac{1}{3}\bar{J}_4\left(\eta_1 + 2\eta_2\left(\bar{I}_1^2 - 2\bar{I}_2\right)\right)\delta_{mn}\right) \tag{4.24}$$

The corresponding elasticity tensor in the reference is obtained as well:

$$\mathbb{C}^{rate-dep} = 4\frac{\partial^2 W}{\partial \dot{\boldsymbol{C}}^2} \tag{4.25}$$

The elasticity tensor is then transformed into the current configuration elasticity tensor:

$$\mathbb{C}_{mnrs}^{rate-dep} = \frac{1}{J} F_{mi} F_{nj} F_{rk} F_{sl} \mathbb{C}_{ijkl}^{rate-dep} \tag{4.26}$$

$$\begin{aligned}\mathbb{C}_{mnrs}^{rate-dep} = &\frac{1}{J}\left(\bar{I}_1 - 3\right)\left(\frac{1}{2}\eta_1\left(\bar{B}_{mr}\bar{B}_{ns} + \bar{B}_{ms}\bar{B}_{nr}\right) + 2\eta_2 \bar{B}_{mk}\bar{B}_{nk}\bar{B}_{rl}\bar{B}_{sl}\right.\\ &-\frac{1}{3}\left(\eta_1 + 2\eta_2\left(\bar{I}_1^2 - 2\bar{I}_2\right)\right)\left(\bar{B}_{mk}\bar{B}_{nk}\delta_{rs} + \bar{B}_{rk}\bar{B}_{sk}\delta_{mn}\right)\\ &\left.+\frac{1}{9}\left(\eta_1\left(\bar{I}_1^2 - 2\bar{I}_2\right) + 2\eta_2\left(\bar{I}_1^2 - 2\bar{I}_2\right)^2\right)\delta_{mn}\delta_{rs}\right)\end{aligned} \tag{4.27}$$

In Section 3.9 we introduced an example of an anisotropic dielectric elastomer. Here, we update the Equation (3.113)–(3.117) to include rate-dependency characteristics for the dielectric elastomer, as well.

$$W = \bar{W} + W^{\text{vol}} + W^{\text{dielectric}} + W^{\text{aniso}} + \bar{W}^{\text{rate-dep}} + W^{\text{rate-dep}} \tag{4.28}$$

$$\begin{aligned}W = &C_{10}\left(\bar{I}_1 - 3\right) + C_{01}\left(\bar{I}_2 - 3\right) + \frac{\kappa}{2}(J-1)^2 - \frac{1}{2}\varepsilon J C^{-1}:\left(\mathbb{E}\otimes\mathbb{E}\right)\\ &+\frac{k_1}{2k_2}\sum_{i=4,6}\left\{\exp\left[k_2\left(\bar{I}_i - 1\right)^2\right] - 1\right\} + \frac{1}{4}\left(\bar{I}_1 - 3\right)\left(\eta_1 \bar{J}_2 + \eta_2 \bar{J}_4^2\right)\end{aligned} \tag{4.29}$$

$$\begin{aligned}S_{ij} = &2\left(C_{10}J^{-\frac{2}{3}}\left(\delta_{ij} - \frac{1}{3}I_1 C_{ij}^{-1}\right) + C_{01}J^{-\frac{4}{3}}\left(I_1\delta_{ij} - C_{ij} - \frac{2}{3}I_2 C_{ij}^{-1}\right)\right.\\ &\left.+\frac{\kappa}{2}(J-1)JC_{ij}^{-1}\right) - \frac{1}{2}\varepsilon J \mathbb{E}_p \mathbb{E}_q\left(C_{ij}^{-1}C_{pq}^{-1} - \frac{1}{2}\left(C_{pj}^{-1}C_{qi}^{-1} + C_{pi}^{-1}C_{qj}^{-1}\right)\right)\\ &+2k_1^{(1)}\left(I_4 - 1\right)\left(\exp\left(k_2^{(1)}\left(I_4 - 1\right)^2\right)\right)A_i^{(1)}A_j^{(1)}\\ &+2k_1^{(2)}\left(I_6 - 1\right)\left(\exp\left(k_2^{(2)}\left(I_6 - 1\right)^2\right)\right)A_i^{(2)}A_j^{(2)}\\ &+\frac{1}{2}\left(\bar{I}_1 - 3\right)\left(\eta_1 J^{-\frac{2}{3}}\dot{\bar{C}}_{ij} 2\eta_2 \bar{J}_4 J^{-\frac{4}{3}} C_{ij} - \frac{1}{3}\bar{J}_4\left(\eta_1 + 2\eta_2\left(\bar{I}_1^2 - 2\bar{I}_2\right)\right)C_{ij}^{-1}\right)\end{aligned} \tag{4.30}$$

$$\sigma_{mn} = \frac{2}{J}\left( C_{10}\left( \bar{B}_{mn} - \frac{1}{3}\bar{I}_1\delta_{mn} \right) + C_{01}\left( \bar{I}_1\bar{B}_{mn} - \bar{B}_{mk}\bar{B}_{nk} - \frac{2}{3}\bar{I}_2\delta_{mn} \right) \right.$$
$$+\frac{\kappa}{2}(J-1)J\delta_{mn}\Bigg) + \varepsilon\left( \mathrm{e}_m\mathrm{e}_n - \frac{1}{2}\mathrm{e}_l\mathrm{e}_l\delta_{mn} \right)$$
$$+\frac{2}{J}\left( k_1^{(1)}(I_4-1)\left(\exp\left(k_2^{(1)}(I_4-1)^2\right)\right)a_m^{(1)}a_n^{(1)} \right.$$
$$\left. +k_1^{(2)}(I_6-1)\left(\exp\left(k_2^{(2)}(I_6-1)^2\right)\right)a_m^{(2)}a_n^{(2)} \right)$$
$$+\frac{1}{2J}(\bar{I}_1-3)\left( 2\eta_1\bar{B}_{mi}D_{ij}\bar{B}_{jn} + 2\left( \eta_2\bar{J}_4 - \frac{1}{3}\eta_1\mathrm{tr}(d) \right)\bar{B}_{mk}\bar{B}_{nk} \right.$$
$$\left. -\frac{1}{3}\bar{J}_4\left(\eta_1 + 2\eta_2\left(\bar{I}_1^2 - 2\bar{I}_2\right)\right)\delta_{mn} \right) \tag{4.31}$$

$$\mathbb{C}_{mnrs} = \left( \frac{4}{3}\left(C_{10}\bar{I}_1 + 4C_{01}\bar{I}_2\right) + \kappa(2J-1) \right)\delta_{mn}\delta_{rs}$$
$$+\left(2\bar{I}_1C_{10} + 4\bar{I}_2C_{01} + \kappa(1-J)\right)\left(\delta_{mr}\delta_{ns} + \delta_{nr}\delta_{ms}\right)$$
$$-4\left(C_{10} + 2C_{01}\bar{I}_1\right)\left(\bar{B}_{mn}\delta_{rs} + \bar{B}_{rs}\delta_{mn}\right)$$
$$+4C_{01}\left(2\left(\bar{B}_{rk}\bar{B}_{sk}\delta_{mn} + \bar{B}_{mk}\bar{B}_{nk}\delta_{rs}\right) - 2\bar{B}_{mn}\bar{B}_{rs} + \left(\bar{B}_{mr}\bar{B}_{ns} + \bar{B}_{nr}\bar{B}_{ms}\right)\right)$$
$$-\frac{1}{2}\varepsilon\left(\left(\delta_{mn}\delta_{rs} - \left(\delta_{mr}\delta_{ns} + \delta_{ms}\delta_{nr}\right)\right)\mathrm{e}_l\mathrm{e}_l\right)$$
$$+\varepsilon\left(\mathrm{e}_n\mathrm{e}_m\delta_{rs} + \mathrm{e}_r\mathrm{e}_s\delta_{mn} - \mathrm{e}_n\mathrm{e}_s\delta_{mr} - \mathrm{e}_m\mathrm{e}_r\delta_{ns} - \mathrm{e}_m\mathrm{e}_s\delta_{nk} - \mathrm{e}_n\mathrm{e}_r\delta_{ms}\right)$$
$$+\frac{4}{J}\left( k_1^{(1)}\left(1 + 2k_2^{(1)}(I_4-1)^2\right)\exp\left(k_2^{(1)}(I_4-1)^2\right)a_m^{(1)}a_n^{(1)}a_r^{(1)}a_s^{(1)} \right.$$
$$\left. +k_1^{(2)}\left(1 + 2k_2^{(2)}(I_6-1)^2\right)\exp\left(k_2^{(2)}(I_6-1)^2\right)a_m^{(2)}a_n^{(2)}a_r^{(2)}a_s^{(2)} \right)$$
$$+\frac{1}{J}(\bar{I}_1-3)\left( \frac{1}{2}\eta_1\left(\bar{B}_{mr}\bar{B}_{ns} + \bar{B}_{ms}\bar{B}_{nr}\right) + 2\eta_2\bar{B}_{mk}\bar{B}_{nk}\bar{B}_{rl}\bar{B}_{sl} \right.$$
$$-\frac{1}{3}J^{-\frac{4}{3}}\left(\eta_1 + 2\eta_2\left(\bar{I}_1^2 - 2\bar{I}_2\right)\right)\left(\bar{B}_{mk}\bar{B}_{nk}\delta_{rs} + \bar{B}_{rk}\bar{B}_{sk}\delta_{mn}\right)$$
$$\left. +\frac{1}{9}\left( \eta_1\left(\bar{I}_1^2 - 2\bar{I}_2\right) + 2\eta_2\left(\bar{I}_1^2 - 2\bar{I}_2\right)^2 \right)\delta_{mn}\delta_{rs} \right) \tag{4.32}$$

## 4.3 VISCOELASTIC EFFECTS

Many materials in engineering and physics are not elastic, so the mentioned hyperelastic constitutive laws are not enough to describe the behavior of these materials because any acceptable process for these materials is associated with energy loss. Appropriate models, categorized in any of differential operator based or convolution integrals base, also describe properties including creep (increasing strain due to constant stress), stress relaxation (decreasing stress due to constant strain), Mullins effect (progressive cyclic softening), hysteresis effect, frequency and rate dependency. In the following, the study of inelastic materials and the explanation of the constitutive law used for visco-hyperelastic materials using convolution integral will be discussed.

In this section, we express the linear response of elastomers using a generalized Maxwell rheological model. Then the response is derived using a general three-dimensional formulation according to Kaliske and Rothert [4]. For this purpose, a one-dimensional modeling framework for small strain is generalized to a three-dimensional one to simulate the finite elements analysis of the elastomers.

## 4.4 RHEOLOGICAL MODELS

Applying models is an efficient method to investigate the mechanical behavior of viscoelastic materials. Since the stress of a perfectly elastic material is proportional to its strain, rheological models estimate a Hookean material with an elastic spring. Furthermore, the linear relation between the stress and strain rate in a perfect Newtonian fluid is modeled using a dashpot. A viscoelastic material behaves between these two extreme cases. Therefore, its behavior would be captured using a combination of springs and dashpot elements. The number and arrangements of the elements depend on the material's feature. A well-known rheological model is the Maxwell element, which we use to devise the constitutive equations and is depicted in Figure 4.1.

The Maxwell element includes a spring and a dashpot connected in series. Deformations of the spring and dashpot are stated by $\epsilon_e$ and $\epsilon_v$, respectively, and lead to the total elongation of $\epsilon = \epsilon_e + \epsilon_v$ for the whole Maxwell element. Considering $\mu_0$ as the elastic constant of the spring, the stress-strain relation would be $\sigma = \mu_0 \epsilon_e$. The equal stress of the dashpot with the coefficient of viscosity $\eta_0$ is

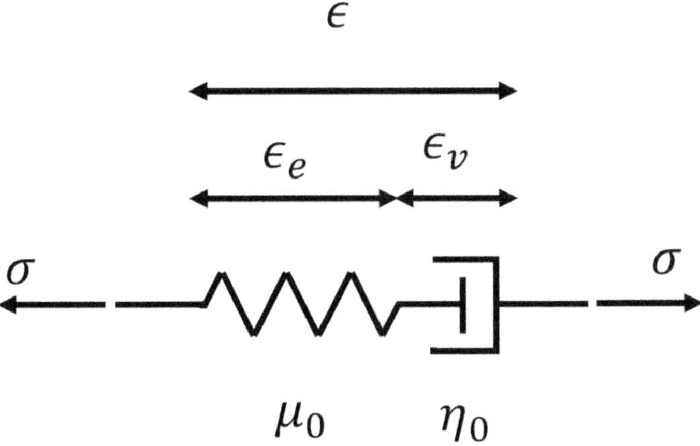


**FIGURE 4.1** Maxwell element.

related to the dashpot strain rate through $\sigma = \eta_0 \dot{\epsilon}_v$. Using the total $\dot{\epsilon}$, we derive the stress-strain relationship in this model as follows:

$$\dot{\sigma} + \frac{1}{\tau_0}\sigma = \mu_0 \dot{\epsilon} \tag{4.33}$$

where $\tau_0 = \frac{\eta_0}{\mu_0}$ represents the relaxation time of the model. The relaxation time characterizes viscoelastic quality of the material by determination of the needed time for the stress to decay. Equation (4.33) has a homogeneous solution:

$$\sigma_h = C_0 \exp\left(-\frac{t}{\tau_0}\right) \tag{4.34}$$

The solution depends on the initial conditions to determine the constant value of $C_0$. Conducting a stress relaxation test, i.e. $\epsilon(0) = \epsilon(t) =$ constant leads to the stress response shown in Figure 4.2, according to a Maxwell element.

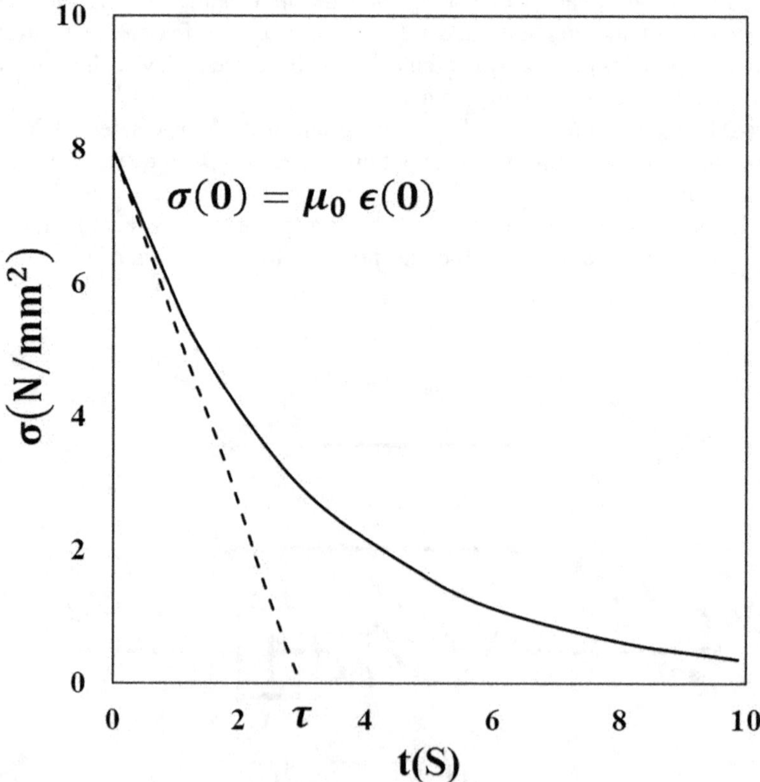


FIGURE 4.2 Stress relaxation test response for a Maxwell element.

According to Figure 4.2, the elastic response defines the initial stress of the material since at time $t = 0$, the stress is $\sigma(0) = \mu_0 \epsilon(0)$. It is used to determine $C_0$.

Taking the essentials of stress relaxation test in account (constant strain), the strain rate would be zero and it declares the particular solution of Equation (4.33): $\sigma_p = 0$. All in all, the total solution would be:

$$\sigma(t) = \mu_0 \exp\left(-\frac{t}{\tau_0}\right)\epsilon(0) \tag{4.35}$$

The stress relaxation function is defined as follows:

$$\Gamma(t) = \mu_0 \exp\left(-\frac{t}{\tau_0}\right) \tag{4.36}$$

This indicates the viscoelastic properties of the material. This answer can be generalized to a generalized Maxwell model in Figure 4.3. The generalized Maxwell model comprises a Hookean spring parallel to some Maxwell elements.

The modulus of the single spring element parallel to the Maxwell elements is $\mu_0$. The modulus of the springs and viscosities of the dampers in Maxwell elements are represented as $\mu_i$ and $\eta_i, i = 1,2,3,\ldots N'$ , respectively.

According to the governing equations of the consisting elements and their configurations, the total stress of the generalized model, $\sigma$, is equal to the sum of the stresses on the single spring and parallel Maxwell elements:

$$\sigma(t) = \mu_0 \epsilon(t) + \sum_{i=1}^{N'} \mu_i \exp\left(-\frac{t}{\tau_i}\right)\epsilon(0) = \Gamma(t)\epsilon(0) \tag{4.37}$$

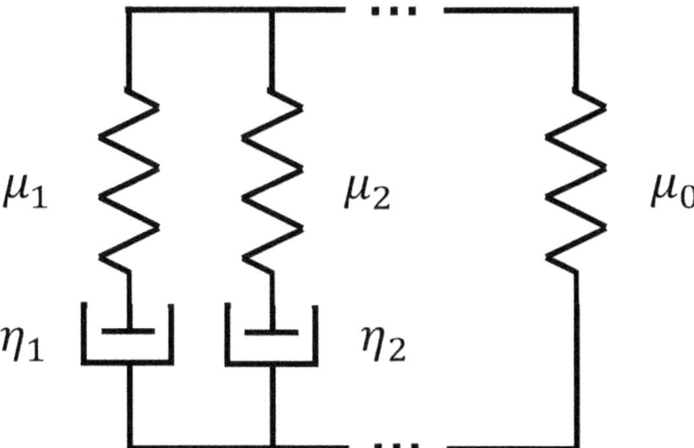


**FIGURE 4.3** Generalized Maxwell model.

$\tau_i$ represents the relaxation time of the ith Maxwell element. It is equal to the constant ratio of the damper's viscosity to the spring modulus in the ith Maxwell element. Therefore, the relaxation function of the generalized model is defined as:

$$\Gamma(t) = \mu_0 + \sum_{i=1}^{N'} \mu_i \exp\left(-\frac{t}{\tau_i}\right) \tag{4.38}$$

To avoid unnecessary complexity, the normalized form of the relaxation function, using the definition of $\gamma_i = \dfrac{\mu_i}{\mu_0}$ is suggested as follows:

$$\gamma(t) = \frac{\Gamma(t)}{\mu_0} = 1 + \sum_{i=1}^{N'} \gamma_i \exp\left(-\frac{t}{\tau_i}\right) \tag{4.39}$$

The Cauchy stress for incremental strains, $\Delta\epsilon_i$ is given below:

$$\sigma(t) = \int_0^t \Gamma(t-s)\frac{\partial \epsilon}{\partial s} ds \tag{4.40}$$

where the relaxation function $\Gamma(t-s)$ is rewritten as follows

$$\Gamma(t-s) = \mu_0 + \sum_{i=1}^{N'} \mu_i \exp\left(-\frac{t-s}{\tau_i}\right) \tag{4.41}$$

Replacing (4.41) in (4.42), the stress expression would expand to:

$$\sigma(t) = \int_0^t \mu_0 \frac{\partial \epsilon(s)}{\partial s} ds + \int_0^t \sum_{i=1}^{N'} \mu_i \exp\left(-\frac{t-s}{\tau_i}\right)\frac{\partial \epsilon(s)}{\partial s} ds \tag{4.42}$$

$\mu_0$ is constant. Also, we know that $\int_0^t \frac{\partial \epsilon(s)}{\partial s} ds = (t)$. Therefore, Equation (4.42) would transform to:

$$\sigma(t) = \mu_0 \epsilon(t) + \int_0^t \sum_{i=1}^{N'} \mu_i \exp\left(-\frac{t-s}{\tau_i}\right)\frac{\partial \epsilon(s)}{\partial s} ds \tag{4.43}$$

Considering $\sigma_0(t) = \mu_0 \epsilon(t)$ as the stress of the elastic element (the Hookean spring) and $h_i(t)$, as the internal stresses in the Maxwell elements leads to:

$$\sigma(t) = \sigma_0(t) + \sum_{i=1}^{N'} h_i(t) \tag{4.44}$$

$h_i(t)$ *is*:

$$h_i(t) = \int_0^t \mu_i \exp\left(-\frac{t-s}{\tau_i}\right)\frac{\partial \epsilon(s)}{\partial s}ds \tag{4.45}$$

Substituting Equations (4.43) and (4.44) and $\epsilon(s) = \sigma_0(s)/\mu_0$ Equation (4.45) can be expressed as:

$$h_i(t) = \int_0^t \gamma_i \exp\left(-\frac{t-s}{\tau_i}\right)\frac{\partial \sigma_0(s)}{\partial s}ds \tag{4.46}$$

For infinite times, $h_i(t)$, the internal stress variables reach zero. Therefore, the stress of the material under a constant deformation decreases to the value of elastic stress.

To determine $h_i(t)$, we apply a finite difference method to Equation (4.46). For a defined time interval, $\left[t_n, t_{n+1}\right]$, the deformation is kept constant for each time increment from $t_n$ to $t_{n+1}$. Therefore, stress is only a function of time. Separation of exponential expressions using the time step $\Delta t = t_{n+1} - t_n$ and integrating Equation (4.46) from zero up to time $t_{n+1}$ gives the result below:

$$h_i(t_{n+1}) = \gamma_i \int_0^{t_{n+1}} \exp\left(-\frac{t_{n+1}-s}{\tau_i}\right)\frac{\partial \sigma_0(s)}{\partial s}ds \tag{4.47}$$

Equation (4.47) is reformed to (4.48) using the separation of deformation history into two stages, $0 \le s \le t_n$ and $t_n \le s \le t_{n+1}$:

$$h_i(t_{n+1}) = \gamma_i \int_0^{t_n} \exp\left(-\frac{t_{n+1}-s}{\tau_i}\right)\frac{\partial \sigma_0(s)}{\partial s}ds + \gamma_i \int_{t_n}^{t_{n+1}} \exp\left(-\frac{t_{n+1}-s}{\tau_i}\right)\frac{\partial \sigma_0(s)}{\partial s}ds \tag{4.48}$$

Using $\Delta t = t_{n+1} - t_n$, Equation (4.48) is changed to

$$\begin{aligned} h_i(t_{n+1}) = {} & \exp\left(-\frac{\Delta t}{\tau_i}\right)\gamma_i \int_0^{t_n} \exp\left(-\frac{t_n-s}{\tau_i}\right)\frac{\partial \sigma_0(s)}{\partial s}ds \\ & + \gamma_i \int_{t_n}^{t_{n+1}} \exp\left(-\frac{t_{n+1}-s}{\tau_i}\right)\frac{\partial \sigma_0(s)}{\partial s}ds \end{aligned} \tag{4.49}$$

By comparing Equations (4.49) and (4.46), we conclude:

$$h_i(t_n) = \gamma_i \int_0^{t_n} \exp\left(-\frac{t_n-s}{\tau_i}\right)\frac{\partial \sigma_0(s)}{\partial s}ds \tag{4.50}$$

Then Equation (4.49) would be:

$$h_i\left(t_{n+1}\right)=\exp\left(-\frac{\Delta t}{\tau_i}\right)h_i\left(t_n\right)+\gamma_i\int_{t_n}^{t_{n+1}}\exp\left(-\frac{t_{n+1}-s}{\tau_i}\right)\frac{\partial\sigma_0\left(s\right)}{\partial s}ds \tag{4.51}$$

Application of a similar finite difference approach for $\frac{d\sigma_0\left(s\right)}{ds}$ simplifies the integral in Equation (4.51) to:

$$\frac{d\sigma_0\left(s\right)}{ds}=\lim_{\Delta s\to 0}\left(\frac{\Delta\sigma_0\left(s\right)}{\Delta s}\right)=\lim_{\Delta s\to 0}\left(\frac{\sigma_0^{n+1}-\sigma_0^n}{\Delta s}\right) \tag{4.52}$$

Adoption of the second-order discrete time approximation in the numerical solution formula, Equation (4.47) would change to:

$$h_i^{n+1}=\exp\left(-\frac{\Delta t}{\tau_i}\right)h_i^n+\frac{\gamma_i\left(1-\exp\left(-\frac{\Delta t}{\tau_i}\right)\right)}{\frac{\Delta t}{\tau_i}}\left(\sigma_0^{n+1}-\sigma_0^n\right) \tag{4.53}$$

Equation (4.53) is an iterative formulation. Therefore, all the values of $h_i^{n+1}$ depends on the previous values of $h_i^n$. Knowing values of $h_i^n$, one can obtain the values for $h_i^{n+1}$.

Then the corresponding iterative expression of (4.43) becomes:

$$\sigma^{n+1}=\sigma_0^{n+1}+\sum_{i=1}^{N'}h_i^{n+1} \tag{4.54}$$

The above stress response shows a one-dimensional numerical formula for a linear response of viscoelastomers. It can be seen that the stress relaxation function is replaced by an exponential series that depends on the spring and damper modules in the model. To introduce the three-dimensional formula of tensorial writing, Equation (4.54) is adopted:

$$\sigma_{ij}^{n+1}=\sigma_{0ij}^{n+1}+\sum_{i=1}^{N'}h_i^{n+1} \tag{4.55}$$

The elastic stress in Equation (4.55) is defined as follows:

$$\sigma_{0ij}^{n+1}=\mathbb{C}_{ijkl}^{e}\epsilon_{kl}^{n+1} \tag{4.56}$$

Here, $\mathbb{C}^e_{ijkl}$ is a fourth-order tensor, known as the elasticity tensor. This tensor describes an elastic relationship between stress and strain.

Using $\mathbb{C}^e_{ijkl}$, the internal stress variables in Equation (4.53) reform to:

$$h_i^{n+1} = \exp\left(-\frac{\Delta t}{\tau_i}\right)h_i^n + \frac{\gamma_i\left(1-\exp\left(-\frac{\Delta t}{\tau_i}\right)\right)}{\frac{\Delta t}{\tau_i}}\left(\mathbb{C}^e_{MK}\epsilon_K^{n+1} - \mathbb{C}^e_{MK}\epsilon_K^n\right) \tag{4.57}$$

The incremental formula for Cauchy stress is obtained as well:

$$\sigma_M^{n+1} = \mathbb{C}^e_{MK}\epsilon_K^n + \sum_{j=1}^{N'} \exp\left(-\frac{\Delta t}{\tau_j}\right)h_j^n + \gamma_j\tau_j\frac{\left(1-\exp\left(-\frac{\Delta t}{\tau_j}\right)\right)}{\Delta t}\left(\mathbb{C}^e_{MK}\epsilon_K^{n+1} - \mathbb{C}^e_{MK}\epsilon_K^n\right) \tag{4.58}$$

Finally, the iterative formula for the elasticity tensor would be:

$$\mathbb{C}^{n+1} = \frac{\partial\Delta\sigma}{\partial\Delta\epsilon} = \left\{1 + \sum_{j=1}^{3}\gamma_j\tau_j\frac{\left(1-\exp\left(-\frac{\Delta t}{\tau_j}\right)\right)}{\Delta t}\right\}\mathbb{C}^n \tag{4.59}$$

## 4.5 CONCLUSION

This chapter presented two different methods to model viscoelastic properties of dielectric elastomers. The first method considers an additive term to the strain energy function to include rate-dependency. Then, successive steps of derivation with respect to the right Cauchy-Green strain rate tensor, develops the constitutive law that captures rate-dependency in addition to hyperelasticity.

The second method makes use of convolution integral to model liner visco-elasticity at small and large deformations. It employs the generalized Maxwell element model to capture the time-dependent behavior of dielectric elastomer materials.

Both methods are used to extend the presented constitutive laws in the previous chapter. The extended model could produce an efficient numerical formulation. The numerical implementations will be employed in finite element approaches to analyze large-scale and complicated cases.

## REFERENCES

[1] K. Upadhyay, G. Subhash, D. Spearot, Visco-hyperelastic constitutive modeling of strain rate sensitive soft materials, *Journal of the Mechanics and Physics of Solids*, 135 (2020) 103777.

[2] D.P. Pioletti, L. Rakotomanana, J.-F. Benvenuti, P.-F. Leyvraz, Viscoelastic constitutive law in large deformations: application to human knee ligaments and tendons, *Journal of Biomechanics*, 31(8) (1998) 753–757.

[3] Z. M. Ghahfarokhi, M. Salmani-Tehrani, M. M. Zand, S. Esmaeilian, A new viscous potential function for developing the viscohyperelastic constitutive model for bovine liver tissue: Continuum formulation and finite element implementation, *International Journal of Applied Mechanics*, 12(03) (2020) 2050029.

[4] M. Kaliske, H. Rothert, Formulation and implementation of three-dimensional viscoelasticity at small and finite strains, *Computational Mechanics*, 19(3) (1997) 228–239.

# 5 Transient and Dynamic Modeling

## 5.1 DYNAMIC ANALYSIS

The dynamic response of dielectric elastomers is observed to be nonlinear [1, 2] and dependent on the applied electric excitation [3]. These features can address some typical issues for traditional devices. For instance, some researchers [2, 4, 5] have reported tunable frequencies for dielectric elastomer membrane which solve the problem of fixed resonance frequency in traditional materials. Such dynamic features of dielectric elastomers revealed a potential to design devices e.g. loudspeakers [6–8], pumps [9–11] and resonators [2, 12–14].

Motivated by these applications, researchers have studied nonlinear dynamic performance of DEAs with various geometries including membranes [15, 16], balloons [1, 7, 17], microbeams [18, 19], and spherical shells [20, 21], recently. They mainly utilized the principal of virtual work [22–24] or Euler-Lagrangian equation [25–28] to examine dynamic behavior of DEAs.

Therefore, we review the governing equations for the dynamic and transient behavior of a planar dielectric elastomer actuator.

## 5.2 PROBLEM DEFINITION

We assume a planar dielectric elastomer actuator, reinforced by two families of fibers to deform as an electric field is applied to it. The reference configuration of the actuator is presented in Figure 5.1. Without electrical or mechanical loading, the elastomer film dimensions are $L_1 \times L_2 \times L_3$. Application of voltage deforms the actuator to the current configuration with the new size: $l_1 \times l_2 \times l_3$. In this deformation, principal stretches are defined as $\lambda_3 = l_3 / L_3, \lambda_2 = l_2 / L_2, \lambda_1 = l_1 / L_1$. The actuator is assumed to be incompressible. The incompressibility criteria, $J = \lambda_1 \lambda_2 \lambda_3 = 1$, leads to the relation $\lambda_3 = \lambda_1^{-1} \lambda_2^{-1}$.

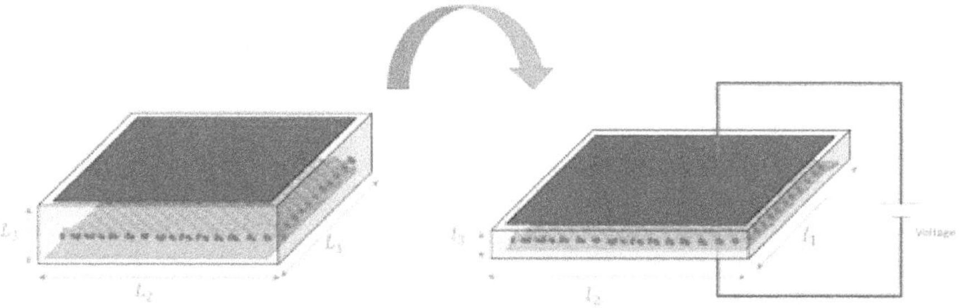

**FIGURE 5.1** Electric actuation.

DOI: 10.1201/9781003571469-7

Therefore, we have:

$$x_1 = \lambda_1(t)X_1$$

$$x_2 = \lambda_2(t)X_2$$

$$x_3 = \lambda_1^{-1}(t)\lambda_2^{-1}(t)X_3 \tag{5.1}$$

$$\dot{x}_1 = \dot{\lambda}_1 X_1$$

$$\dot{x}_2 = \dot{\lambda}_2 X_2$$

$$\dot{x}_3 = -\left(\lambda_1^{-2}\lambda_2^{-1}\dot{\lambda}_1 + \lambda_1^{-1}\lambda_2^{-2}\dot{\lambda}_2\right)X_3 = -\lambda_1^{-1}\lambda_2^{-1}\left(\lambda_1^{-1}\dot{\lambda}_1 + \lambda_2^{-1}\dot{\lambda}_2\right)X_3 \tag{5.2}$$

$$\boldsymbol{F} = \begin{bmatrix} \lambda_1 & 0 & 0 \\ 0 & \lambda_2 & 0 \\ 0 & 0 & \lambda_1^{-1}\lambda_2^{-1} \end{bmatrix} \tag{5.3}$$

$$\boldsymbol{C} = \begin{bmatrix} \lambda_1^2 & 0 & 0 \\ 0 & \lambda_2^2 & 0 \\ 0 & 0 & \lambda_1^{-2}\lambda_2^{-2} \end{bmatrix} \tag{5.4}$$

Considering the mentioned features for the actuator, we take Equation (5.5) to describe its strain energy function.

$$W = W^{\text{hyper}} + W^{\text{dielectric}} + W^{\text{aniso}} = C_{10}\left(I_1 - 3\right) + C_{01}\left(I_2 - 3\right) - \frac{1}{2}\varepsilon J\boldsymbol{C}^{-1} : \left(\mathbb{E} \otimes \mathbb{E}\right)$$
$$+ \frac{k_1}{2k_2}\sum_{i=4,6}\left\{\exp\left[k_2\left(I_i - 1\right)^2\right] - 1\right\} \tag{5.5}$$

To develop the model as a function of the principal stretches, explicitly, we need the following expressions:

$$I_1 = \lambda_1^2 + \lambda_2^2 + \lambda_1^{-2}\lambda_2^{-2} \tag{5.6}$$

$$I_2 = \left(\lambda_1\lambda_2\right)^2 + \lambda_1^{-2} + \lambda_2^{-2} \tag{5.1}$$

$$\boldsymbol{C}^{-1} = \begin{bmatrix} \lambda_1^{-2} & 0 & 0 \\ 0 & \lambda_2^{-2} & 0 \\ 0 & 0 & \lambda_1^2\lambda_2^2 \end{bmatrix} \tag{5.7}$$

$$\mathbb{E} = \begin{bmatrix} E_1 \\ E_2 \\ E_3 \end{bmatrix} \tag{5.8}$$

$$\mathbb{E} \otimes \mathbb{E} = \begin{bmatrix} E_1^2 & E_1E_2 & E_1E_3 \\ E_1E_2 & E_2^2 & E_2E_3 \\ E_1E_3 & E_2E_3 & E_3^2 \end{bmatrix} \tag{5.9}$$

$$\boldsymbol{C}^{-1} : \left(\mathbb{E} \otimes \mathbb{E}\right) = \lambda_1^{-2}E_1^2 + \lambda_2^{-2}E_2^2 + \lambda_1^2\lambda_2^2E_3^2 \tag{5.10}$$

Tensors and parameters related to anisotropy are also required. They are defined according to the configuration demonstrated in Figure 5.2.

$$\boldsymbol{A}^{(1)} = \begin{bmatrix} \cos\alpha \\ \sin\alpha \\ 0 \end{bmatrix}$$

$$\boldsymbol{A}^{(2)} = \begin{bmatrix} \cos-\alpha \\ \sin-\alpha \\ 0 \end{bmatrix} \tag{5.11}$$

$$I_4 = I_6 = \boldsymbol{A}^{(1)} \cdot \boldsymbol{C}\boldsymbol{A}^{(1)} = \lambda_1^2 \cos^2\alpha + \lambda_2^2 \sin^2\alpha \tag{5.12}$$

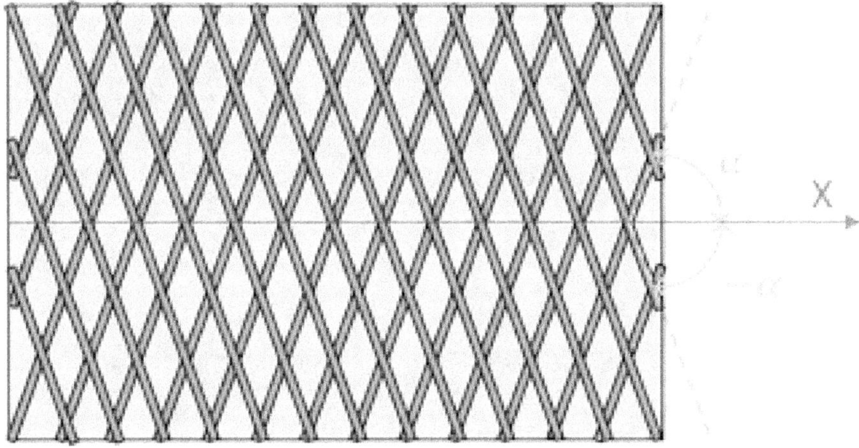


**FIGURE 5.2** Fiber families configuration.

Therefore, the final form of strain energy function would be

$$
\begin{aligned}
W &= C_{10}\left(\lambda_1^2+\lambda_2^2+\lambda_1^{-2}\lambda_2^{-2}-3\right)+C_{01}\left(\lambda_1^2+\lambda_2^2+\lambda_1^{-2}+\lambda_2^{-2}-3\right) \\
&-\frac{1}{2}\varepsilon\left(\lambda_1^{-2}E_1^2+\lambda_2^{-2}E_2^2+\lambda_1^2\lambda_2^2E_3^2\right) \\
&+\frac{k_1}{k_2}\left(\exp\left[k_2\left(\lambda_1^2\cos^2\alpha+\lambda_2^2\sin^2\alpha-1\right)^2\right]-1\right)
\end{aligned}
\tag{5.13}
$$

## 5.3 GOVERNING EQUATION

We assume the actuator to be a non-conservative system. Then, the Euler-Lagrangian equation is used to extract the equation of motion:

$$
\frac{\partial}{\partial t}\left(\frac{\partial \mathcal{L}}{\partial \dot{\lambda}_i}\right)-\frac{\partial \mathcal{L}}{\partial \lambda_i}=0, i=1,2 \tag{5.14}
$$

where $i$ is determined according to the number of independent generalized coordinates. $\mathcal{L}$ denotes the lagrangian. It is defined based on the kinetic and potential energies, $K$ and $U$, respectively:

$$
\mathcal{L}=K-U \tag{5.15}
$$

In absence of mechanical loads, potential energy of the system consists of elastic energy and electrostatic energy. So, it is obtained by integrating the strain energy density on the volume of the actuator. It is worth noting that integrating takes place in the current configuration:

$$
\begin{aligned}
U = \int_V W dV &= L_1L_2L_3\Big\{C_{10}\left(\lambda_1^2+\lambda_2^2+\lambda_1^{-2}\lambda_2^{-2}-3\right)+C_{01}\left(\lambda_1^2+\lambda_2^2+\lambda_1^{-2}+\lambda_2^{-2}-3\right) \\
&-\frac{1}{2}\varepsilon\left(\lambda_1^{-2}E_1^2+\lambda_2^{-2}E_2^2+\lambda_1^2\lambda_2^2E_3^2\right) \\
&+\frac{k_1}{k_2}\left(\exp\left[k_2\left(\lambda_1^2\cos^2\alpha+\lambda_2^2\sin^2\alpha-1\right)^2\right]-1\right)\Big\} \\
&= V\Big\{C_{10}\left(\lambda_1^2+\lambda_2^2+\lambda_1^{-2}\lambda_2^{-2}-3\right)+C_{01}\left(\lambda_1^2+\lambda_2^2+\lambda_1^{-2}+\lambda_2^{-2}-3\right) \\
&-\frac{1}{2}\varepsilon\left(\lambda_1^{-2}E_1^2+\lambda_2^{-2}E_2^2+\lambda_1^2\lambda_2^2E_3^2\right) \\
&+\frac{k_1}{k_2}\left(\exp\left[k_2\left(\lambda_1^2\cos^2\alpha+\lambda_2^2\sin^2\alpha-1\right)^2\right]-1\right)\Big\}
\end{aligned}
\tag{5.16}
$$

The kinetic energy of the system is derived using the following expression:

$$
\begin{aligned}
K &= \frac{1}{2}\rho\int_0^{L_3}\int_0^{L_2}\int_0^{L_1}\left(\dot{x}_1^2+\dot{x}_2^2+\dot{x}_3^2\right)dX_1dX_2dX_3 \\
&= \frac{1}{2}\rho\int_0^{L_3}\int_0^{L_2}\int_0^{L_1}\dot{\lambda}_1^2X_1^2dX_1dX_2dX_3+\frac{1}{2}\rho\int_0^{L_3}\int_0^{L_2}\int_0^{L_1}\dot{\lambda}_2^2X_2^2dX_1dX_2dX_3 \\
&+\frac{1}{2}\rho\int_0^{L_3}\int_0^{L_2}\int_0^{L_1}\left(\lambda_1^{-1}\lambda_2^{-1}\left(\lambda_1^{-1}\dot{\lambda}_1+\lambda_2^{-1}\dot{\lambda}_2\right)\right)^2X_3^2dX_1dX_2dX_3 \\
&= \frac{1}{2}\rho\dot{\lambda}_1^2L_2L_3\int_0^{L_1}X_1^2dX_1+\frac{1}{2}\rho\dot{\lambda}_2^2L_1L_3\int_0^{L_2}X_2^2dX_2 \\
&+\frac{1}{2}\rho\left(\lambda_1^{-1}\lambda_2^{-1}\left(\lambda_1^{-1}\dot{\lambda}_1+\lambda_2^{-1}\dot{\lambda}_2\right)\right)^2L_1L_2\int_0^{L_3}X_3^2dX_3 \\
&= \frac{1}{6}\rho V\left(l_1^2+l_2^2+l_3^2\left(\lambda_1^{-1}\dot{\lambda}_1+\lambda_2^{-1}\dot{\lambda}_2\right)^2\right)
\end{aligned}
\tag{5.17}
$$

Combining (5.15) and (5.16) and (5.17) leads to the Lagrangian:

$$
\begin{aligned}
\mathcal{L} = K-U &= \frac{1}{6}\rho V\left(l_1^2+l_2^2+l_3^2\left(\lambda_1^{-1}\dot{\lambda}_1+\lambda_2^{-1}\dot{\lambda}_2\right)^2\right) \\
&-V\Bigg\{C_{10}\left(\lambda_1^2+\lambda_2^2+\lambda_1^{-2}\lambda_2^{-2}-3\right)+C_{01}\left(\lambda_1^2+\lambda_2^2+\lambda_1^{-2}+\lambda_2^{-2}-3\right) \\
&-\frac{1}{2}\varepsilon\left(\lambda_1^{-2}E_1^2+\lambda_2^{-2}E_2^2+\lambda_1^2\lambda_2^2E_3^2\right) \\
&+\frac{k_1}{k_2}\left(\exp\left[k_2\left(\lambda_1^2\cos^2\alpha+\lambda_2^2\sin^2\alpha-1\right)^2\right]-1\right)\Bigg\}
\end{aligned}
\tag{5.18}
$$

Referring to (5.14), the following expressions are: required to derive the equation of motion:

$$
\frac{\partial\mathcal{L}}{\partial\dot{\lambda}_1}=\frac{1}{3}\rho Vl_3^2\lambda_1^{-1}\left(\lambda_1^{-1}\dot{\lambda}_1+\lambda_2^{-1}\dot{\lambda}_2\right)
$$

$$
\frac{\partial\mathcal{L}}{\partial\dot{\lambda}_2}=\frac{1}{3}\rho Vl_3^2\lambda_2^{-1}\left(\lambda_1^{-1}\dot{\lambda}_1+\dot{\lambda}_2\dot{\lambda}_2\right)
\tag{5.19}
$$

$$
\begin{aligned}
\frac{\partial}{\partial t}\left(\frac{\partial\mathcal{L}}{\partial\dot{\lambda}_1}\right) &= \frac{1}{3}\rho Vl_3^2\frac{\partial}{\partial t}\left(\lambda_1^{-1}\left(\lambda_1^{-1}\dot{\lambda}_1+\lambda_2^{-1}\dot{\lambda}_2\right)\right) \\
&= \frac{1}{3}\rho Vl_3^2\left(\left(\lambda_1^{-1}\dot{\lambda}_1+\lambda_2^{-1}\dot{\lambda}_2\right)\frac{\partial}{\partial t}\lambda_1^{-1}+\lambda_1^{-1}\frac{\partial}{\partial t}\left(\lambda_1^{-1}\dot{\lambda}_1+\lambda_2^{-1}\dot{\lambda}_2\right)\right) \\
&= \frac{1}{3}\rho Vl_3^2\Big(-\left(\lambda_1^{-1}\dot{\lambda}_1+\lambda_2^{-1}\dot{\lambda}_2\right)\dot{\lambda}_1\lambda_1^{-2} \\
&+\lambda_1^{-1}\left(-\dot{\lambda}_1^2\lambda_1^{-2}+\lambda_1^{-1}\ddot{\lambda}_1-\dot{\lambda}_2^2\lambda_2^{-2}+\lambda_2^{-1}\ddot{\lambda}_2\right)\Big)
\end{aligned}
\tag{5.20}
$$

$$\frac{\partial}{\partial t}\left(\frac{\partial \mathcal{L}}{\partial \dot{\lambda}_2}\right)=\frac{\partial \mathcal{L}}{\partial \dot{\lambda}_2}=\frac{1}{3}\rho V l_3^2 \frac{\partial}{\partial t}\left(\lambda_2^{-1}\left(\lambda_1^{-1}\dot{\lambda}_1+\dot{\lambda}_2\dot{\lambda}_2\right)\right)$$
$$=\frac{1}{3}\rho V l_3^2\left(-\left(\lambda_1^{-1}\dot{\lambda}_1+\lambda_2^{-1}\dot{\lambda}_2\right)\dot{\lambda}_2\lambda_2^{-2}+\lambda_2^{-1}\left(-\dot{\lambda}_1^2\lambda_1^{-2}+\lambda_1^{-1}\ddot{\lambda}_1-\dot{\lambda}_2^2\lambda_2^{-2}+\lambda_2^{-1}\ddot{\lambda}_2\right)\right) \quad (5.21)$$

$$\frac{\partial \mathcal{L}}{\partial \lambda_1}=-\frac{1}{3}\rho V l_3^2\lambda_1^{-2}\dot{\lambda}_1\left(\lambda_1^{-1}\dot{\lambda}_1+\lambda_2^{-1}\dot{\lambda}_2\right)$$
$$-V\left\{2\lambda_1\left(C_{10}+C_{01}\right)-2\lambda_1^{-3}\left(C_{10}\lambda_2^{-2}+C_{01}\right)+\varepsilon\lambda_1^{-1}\left(\lambda_1^{-2}E_1^2-\lambda_1^2\lambda_2^2E_3^2\right)\right.$$
$$\left.+4k_1\lambda_1\cos^2\alpha\left(\lambda_1^2\cos^2\alpha+\lambda_2^2\sin^2\alpha-1\right)\exp\left[k_2\left(\lambda_1^2\cos^2\alpha+\lambda_2^2\sin^2\alpha-1\right)^2\right]\right\}$$
$$(5.22)$$

$$\frac{\partial \mathcal{L}}{\partial \lambda_2}=-\frac{1}{3}\rho V l_3^2\lambda_2^{-2}\dot{\lambda}_2\left(\lambda_1^{-1}\dot{\lambda}_1+\lambda_2^{-1}\dot{\lambda}_2\right)$$
$$-V\left\{2\lambda_2\left(C_{10}+C_{01}\right)-2\lambda_2^{-3}\left(\lambda_1^{-2}C_{10}+C_{01}\right)-\varepsilon\left(-\lambda_2^{-3}E_2^2+\lambda_1^2\lambda_2E_3^2\right)\right.$$
$$\left.+4k_1\lambda_2\sin^2\alpha\left(\lambda_1^2\cos^2\alpha+\lambda_2^2\sin^2\alpha-1\right)\exp\left[k_2\left(\lambda_1^2\cos^2\alpha+\lambda_2^2\sin^2\alpha-1\right)^2\right]\right\}$$
$$(5.23)$$

All in all, equations of motion would be:

$$\frac{\partial}{\partial t}\left(\frac{\partial \mathcal{L}}{\partial \dot{\lambda}_i}\right)-\frac{\partial \mathcal{L}}{\partial \lambda_i}=0, i=1,2 \quad (5.24)$$

$$\frac{1}{3}\rho V l_3^2\lambda_1^{-1}\left(\lambda_1^{-1}\ddot{\lambda}_1+\lambda_2^{-1}\ddot{\lambda}_2-\dot{\lambda}_1^2\lambda_1^{-2}-\dot{\lambda}_2^2\lambda_2^{-2}\right)$$
$$+V\left(2\lambda_1\left(C_{10}+C_{01}\right)-2\lambda_1^{-3}\left(C_{10}\lambda_2^{-2}+C_{01}\right)+\varepsilon\lambda_1^{-1}\left(\lambda_1^{-2}E_1^2-\lambda_1^2\lambda_2^2E_3^2\right)\right.$$
$$+4k_1\lambda_1\cos^2\alpha\left(\lambda_1^2\cos^2\alpha+\lambda_2^2\sin^2\alpha-1\right)$$
$$\left.\exp\left[k_2\left(\lambda_1^2\cos^2\alpha+\lambda_2^2\sin^2\alpha-1\right)^2\right]\right)=0 \quad (5.25)$$

$$\frac{1}{3}\rho V l_3^2\lambda_2^{-1}\left(\lambda_1^{-1}\ddot{\lambda}_1+\lambda_2^{-1}\ddot{\lambda}_2-\dot{\lambda}_1^2\lambda_1^{-2}-\dot{\lambda}_2^2\lambda_2^{-2}\right)$$
$$+V\left(2\lambda_2\left(C_{10}+C_{01}\right)-2\lambda_2^{-3}\left(\lambda_1^{-2}C_{10}+C_{01}\right)+\varepsilon\lambda_2^{-1}\left(\lambda_2^{-2}E_2^2-\lambda_1^2\lambda_2^2E_3^2\right)\right.$$
$$+4k_1\lambda_2\sin^2\alpha\left(\lambda_1^2\cos^2\alpha+\lambda_2^2\sin^2\alpha-1\right)$$
$$\left.\exp\left[k_2\left(\lambda_1^2\cos^2\alpha+\lambda_2^2\sin^2\alpha-1\right)^2\right]\right)=0 \quad (5.26)$$

## 5.4 RATE DEPENDENCY

To include dissipative features, the deformation is assumed to be time-varying. Then, time derivatives are also needed:

$$\dot{C} = \begin{bmatrix} 2\lambda_1\dot{\lambda}_1 & 0 & 0 \\ 0 & 2\lambda_2\dot{\lambda}_2 & 0 \\ 0 & 0 & -2\lambda_1^{-2}\lambda_2^{-2}\left(\lambda_1^{-1}\dot{\lambda}_1 + \lambda_2^{-1}\dot{\lambda}_2\right) \end{bmatrix} \tag{5.27}$$

Moreover, the Lagrangian updates to

$$\frac{\partial}{\partial t}\left(\frac{\partial \mathcal{L}}{\partial \dot{\lambda}_i}\right) - \frac{\partial \mathcal{L}}{\partial \lambda_i} + \frac{\partial \mathcal{D}}{\partial \dot{\lambda}_i} = 0 \tag{5.28}$$

Assuming the dissipation function, $\mathcal{D}$ to be:

$$\mathcal{D} = \frac{1}{4}(I_1 - 3)\left(\eta_1 J_2 + \eta_2 J_4^2\right) \tag{5.29}$$

Additional invariants are required:

$$J_2 = \frac{1}{2}\left(\boldsymbol{I} : \dot{\boldsymbol{C}}^2\right) = 2\left(\dot{\lambda}_1^2\lambda_1^2 + \dot{\lambda}_2^2\lambda_2^2 + \lambda_1^{-4}\lambda_2^{-4}\left(\lambda_1^{-1}\dot{\lambda}_1 + \lambda_2^{-1}\dot{\lambda}_2\right)^2\right) \tag{5.30}$$

$$J_4^2 = 4\left(\dot{\lambda}_1\lambda_1^3 + \dot{\lambda}_2\lambda_2^3 - \lambda_1^{-4}\lambda_2^{-4}\left(\lambda_1^{-1}\dot{\lambda}_1 + \lambda_2^{-1}\dot{\lambda}_2\right)\right)^2 \tag{5.31}$$

Therefore:

$$\begin{aligned} \mathcal{D} &= \frac{1}{4}(I_1 - 3)\left(\eta_1 J_2 + \eta_2 J_4^2\right) \\ &= \frac{1}{4}(I_1 - 3)\left(2\eta_1\left(\dot{\lambda}_1^2\lambda_1^2 + \dot{\lambda}_2^2\lambda_2^2 + \lambda_1^{-4}\lambda_2^{-4}\left(\lambda_1^{-1}\dot{\lambda}_1 + \lambda_2^{-1}\dot{\lambda}_2\right)^2\right)\right. \\ &\left. +4\eta_2\left(\dot{\lambda}_1\lambda_1^3 + \dot{\lambda}_2\lambda_2^3 - \lambda_1^{-4}\lambda_2^{-4}\left(\lambda_1^{-1}\dot{\lambda}_1 + \lambda_2^{-1}\dot{\lambda}_2\right)\right)^2\right) \end{aligned} \tag{5.32}$$

$$\begin{aligned} \frac{\partial \mathcal{D}}{\partial \dot{\lambda}_1} &= (I_1 - 3)\left(\eta_1\left(\dot{\lambda}_1\left(\lambda_1^2 + \lambda_1^{-6}\lambda_2^{-4}\right) + \dot{\lambda}_2\lambda_1^{-5}\lambda_2^{-5}\right)\right. \\ &\left. +\eta_2\left(\dot{\lambda}_1\left(\lambda_1^3 - \lambda_1^{-5}\lambda_2^{-4}\right)^2 + \dot{\lambda}_2\left(\lambda_1^3 - \lambda_1^{-5}\lambda_2^{-4}\right)\left(\lambda_2^3 - \lambda_1^{-4}\lambda_2^{-5}\right)\right)\right) \end{aligned} \tag{5.33}$$

$$\frac{\partial \mathcal{D}}{\partial \dot{\lambda}_2} = (I_1 - 3)\Big(\eta_1\left(\dot{\lambda}_2\left(\lambda_2^2 + \lambda_1^{-4}\lambda_2^{-6}\right) + \dot{\lambda}_1\lambda_1^{-5}\lambda_2^{-5}\right)$$
$$+\eta_2\left(\dot{\lambda}_1\left(\lambda_1^3 - \lambda_1^{-5}\lambda_2^{-4}\right)\left(\lambda_2^3 - \lambda_1^{-4}\lambda_2^{-5}\right) + \dot{\lambda}_2\left(\lambda_2^3 - \lambda_1^{-4}\lambda_2^{-5}\right)^2\right)\Big) \quad (5.34)$$

Finally, the equations of motion are:

$$\frac{1}{3}\rho V l_3^2 \lambda_1^{-1}\left(\lambda_1^{-1}\ddot{\lambda}_1 + \lambda_2^{-1}\ddot{\lambda}_2 - \dot{\lambda}_1^2\lambda_1^{-2} - \dot{\lambda}_2^2\lambda_2^{-2}\right)$$
$$+V\Big(2\lambda_1\left(C_{10} + C_{01}\right) - 2\lambda_1^{-3}\left(C_{10}\lambda_2^{-2} + C_{01}\right) + \varepsilon\lambda_1^{-1}\left(\lambda_1^{-2}E_1^2 - \lambda_1^2\lambda_2^2E_3^2\right)$$
$$+4k_1\lambda_1\cos^2\alpha\left(\lambda_1^2\cos^2\alpha + \lambda_2^2\sin^2\alpha - 1\right)\exp\left[k_2\left(\lambda_1^2\cos^2\alpha + \lambda_2^2\sin^2\alpha - 1\right)^2\right]\Big)$$
$$+(I_1 - 3)\Big(\eta_1\left(\dot{\lambda}_1\left(\lambda_1^2 + \lambda_1^{-6}\lambda_2^{-4}\right) + \dot{\lambda}_2\lambda_1^{-5}\lambda_2^{-5}\right) + \eta_2\Big(\dot{\lambda}_1\left(\lambda_1^3 - \lambda_1^{-5}\lambda_2^{-4}\right)^2$$
$$+\dot{\lambda}_2\left(\lambda_1^3 - \lambda_1^{-5}\lambda_2^{-4}\right)\left(\lambda_2^3 - \lambda_1^{-4}\lambda_2^{-5}\right)\Big)\Big) = 0 \quad (5.35)$$

$$\frac{1}{3}\rho V l_3^2 \lambda_2^{-1}\left(\lambda_1^{-1}\ddot{\lambda}_1 + \lambda_2^{-1}\ddot{\lambda}_2 - \dot{\lambda}_1^2\lambda_1^{-2} - \dot{\lambda}_2^2\lambda_2^{-2}\right)$$
$$+V\Big(2\lambda_2\left(C_{10} + C_{01}\right) - 2\lambda_2^{-3}\left(\lambda_1^{-2}C_{10} + C_{01}\right) + \varepsilon\lambda_2^{-1}\left(\lambda_2^{-2}E_2^2 - \lambda_1^2\lambda_2^2E_3^2\right)$$
$$+4k_1\lambda_2\sin^2\alpha\left(\lambda_1^2\cos^2\alpha + \lambda_2^2\sin^2\alpha - 1\right)\exp\left[k_2\left(\lambda_1^2\cos^2\alpha + \lambda_2^2\sin^2\alpha - 1\right)^2\right]\Big)$$
$$+(I_1 - 3)\Big(\eta_1\left(\dot{\lambda}_2\left(\lambda_2^2 + \lambda_1^{-4}\lambda_2^{-6}\right) + \dot{\lambda}_1\lambda_1^{-5}\lambda_2^{-5}\right)$$
$$+\eta_2\left(\dot{\lambda}_1\left(\lambda_1^3 - \lambda_1^{-5}\lambda_2^{-4}\right)\left(\lambda_2^3 - \lambda_1^{-4}\lambda_2^{-5}\right) + \dot{\lambda}_2\left(\lambda_2^3 - \lambda_1^{-4}\lambda_2^{-5}\right)^2\right)\Big) = 0 \quad (5.36)$$

It worth noting that the electric field applied to the actuator, $\mathbb{E}$ may be expressed as voltage per length or explicit electric field. Also, the electric excitation may be direct or alternative or both. As an example, $E_3 = E_3^{DC} + E_3^{AC}\sin\Omega t$.

To be able to present some semi-analytical solution for the, we go through some common cases, e.g. uniaxial tension and pure shear.

### 5.4.1 Uniaxial Tension

For a uniaxial tension loading, along $X_1$, demonstrated in Figure 5.3, the deformation gradient, right Cauchy-Green strain tensor, corresponding invariants, and $\boldsymbol{C}^{-1}$ will be:

$$\boldsymbol{F} = \begin{bmatrix} \lambda & 0 & 0 \\ 0 & \frac{1}{\sqrt{\lambda}} & 0 \\ 0 & 0 & \frac{1}{\sqrt{\lambda}} \end{bmatrix} \quad (5.37)$$

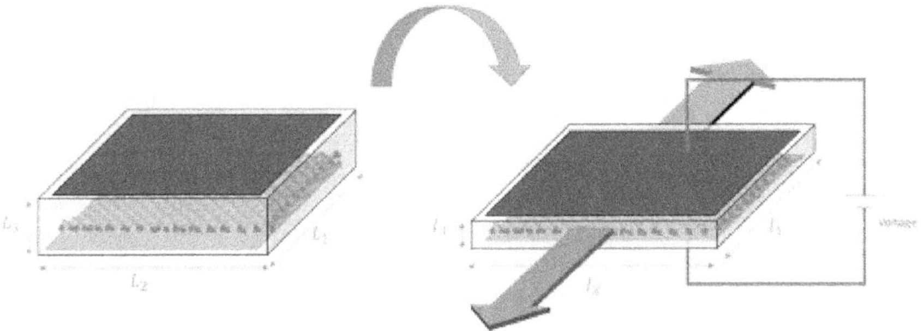

**FIGURE 5.3** Uniaxial tensile loading.

$$\boldsymbol{C} = \begin{bmatrix} \lambda^2 & 0 & 0 \\ 0 & \lambda^{-1} & 0 \\ 0 & 0 & \lambda^{-1} \end{bmatrix} \quad (5.38)$$

$$I_1 = \lambda^2 + 2\lambda^{-1} \quad (5.39)$$

$$I_2 = \lambda^{-2} + 2\lambda \quad (5.40)$$

$$\boldsymbol{C}^{-1} = \begin{bmatrix} \lambda^{-2} & 0 & 0 \\ 0 & \lambda & 0 \\ 0 & 0 & \lambda \end{bmatrix} \quad (5.41)$$

We also assume an electric field along the $X_3$

$$\mathbb{E} = \begin{bmatrix} 0 \\ 0 \\ E_3 \end{bmatrix} \quad (5.42)$$

Then, the needed terms to proceed are:

$$\mathbb{E} \otimes \mathbb{E} = \begin{bmatrix} 0 & 0 & 0 \\ 0 & 0 & 0 \\ 0 & 0 & E_3^2 \end{bmatrix} \quad (5.43)$$

$$\boldsymbol{C}^{-1} : (\mathbb{E} \otimes \mathbb{E}) = \lambda E_3^2 \quad (5.44)$$

On the other hand, the effects of two fiber families are also quantified using:

$$I_4 = I_6 = \boldsymbol{A}^{(1)} \cdot \boldsymbol{C}\boldsymbol{A}^{(1)} = \lambda^2 \cos^2 \alpha + \lambda^{-1} \sin^2 \alpha \quad (5.45)$$

Hence, the potential energy is obtained as:

$$U = \int_V W dV = V\Big\{C_{10}\left(\lambda^2 + 2\lambda^{-1} - 3\right) + C_{01}\left(\lambda^{-2} + 2\lambda - 3\right)$$
$$-\frac{1}{2}\varepsilon\lambda E_3^2 + \frac{k_1}{k_2}\left(\exp\left[k_2\left(\lambda^2\cos^2\alpha + \lambda^{-1}\sin^2\alpha - 1\right)^2\right] - 1\right)\Big\} \tag{5.46}$$

To drive the kinetic energy, velocity components are determined according to the deformation gradient:

$$x_1 = \lambda(t) X_1$$
$$x_2 = \lambda^{-\frac{1}{2}} X_2$$
$$x_3 = \lambda^{-\frac{1}{2}} X_3 \tag{5.47}$$
$$\dot{x}_1 = \dot{\lambda} X_1$$
$$\dot{x}_2 = -\frac{1}{2}\dot{\lambda}\lambda^{-\frac{3}{2}} X_2$$
$$\dot{x}_3 = -\frac{1}{2}\dot{\lambda}\lambda^{-\frac{3}{2}} X_3 \tag{5.48}$$

The final expression for kinetic energy would be:

$$K = \frac{1}{2}\rho\int_0^{L_3}\int_0^{L_2}\int_0^{L_1}\left(\dot{x}_1^2 + \dot{x}_2^2 + \dot{x}_3^2\right) dX_1 dX_2 dX_3$$
$$= \frac{1}{2}\rho\dot{\lambda}^2\int_0^{L_3}\int_0^{L_2}\int_0^{L_1} X_1^2 dX_1 dX_2 dX_3 + \frac{1}{8}\rho\dot{\lambda}^2\lambda^{-3}\int_0^{L_3}\int_0^{L_2}\int_0^{L_1} X_2^2 dX_1 dX_2 dX_3$$
$$+\frac{1}{8}\rho\dot{\lambda}^2\lambda^{-3}\int_0^{L_3}\int_0^{L_2}\int_0^{L_1} X_3^2 dX_1 dX_2 dX_3 = \frac{1}{6}\rho\dot{\lambda}^2 L_1^3 L_2 L_3$$
$$+\frac{1}{24}\rho\dot{\lambda}^2\lambda^{-3} L_2^3 L_1 L_3 + \frac{1}{24}\rho\dot{\lambda}^2\lambda^{-3} L_3^3 L_2 L_1$$
$$= \frac{1}{24}\rho\dot{\lambda}^2 L_1 L_2 L_3\left(4L_1^2 + \lambda^{-3}L_2^2 + \lambda^{-3}L_3^2\right) \tag{5.49}$$

The Lagrangian and related derivatives are as follows:

$$\mathcal{L} = K - U = \frac{1}{24}\rho\dot{\lambda}^2 L_1 L_2 L_3\left(4L_1^2 + \lambda^{-3}L_2^2 + \lambda^{-3}L_3^2\right)$$
$$-V\Big\{C_{10}\left(\lambda^2 + 2\lambda^{-1} - 3\right) + C_{01}\left(\lambda^{-2} + 2\lambda - 3\right)$$
$$-\frac{1}{2}\varepsilon\lambda E_3^2 + \frac{k_1}{k_2}\left(\exp\left[k_2\left(\lambda^2\cos^2\alpha + \lambda^{-1}\sin^2\alpha - 1\right)^2 - 1\right] - 1\right)\Big\} \tag{5.50}$$

$$\frac{\partial\mathcal{L}}{\partial\lambda}=\frac{-1}{8}\rho\dot{\lambda}^2V\left(L_2^2+L_3^2\right)\lambda^{-4}-V\left\{2\left(C_{10}\lambda+C_{01}\right)\left(1-\lambda^{-3}\right)-\frac{1}{2}\varepsilon E_3^2\right\}$$
$$+2k_1\left(\lambda^2\cos^2\alpha+\lambda^{-1}\sin^2\alpha-1\right)\left(2\lambda\cos^2\alpha-\lambda^{-2}\sin^2\alpha\right)$$
$$\exp\left[k_2\left(\lambda^2\cos^2\alpha+\lambda^{-1}\sin^2\alpha-1\right)^2-1\right] \tag{5.51}$$

$$\frac{\partial\mathcal{L}}{\partial\dot{\lambda}}=\frac{1}{12}\rho L_1L_2L_3\left(4L_1^2+\lambda^{-3}L_2^2+\lambda^{-3}L_3^2\right)\dot{\lambda} \tag{5.52}$$

$$\frac{\partial}{\partial t}\left(\frac{\partial\mathcal{L}}{\partial\dot{\lambda}}\right)=\frac{1}{12}\rho L_1L_2L_3\left(4L_1^2+\lambda^{-3}L_2^2+\lambda^{-3}L_3^2\right)\ddot{\lambda} \tag{5.53}$$

Then, the Euler-Lagrange equation to conclude the equation of motion is:

$$\frac{\partial}{\partial t}\left(\frac{\partial\mathcal{L}}{\partial\dot{\lambda}}\right)-\frac{\partial\mathcal{L}}{\partial\lambda}=0 \tag{5.54}$$

$$\frac{1}{12}\rho V\left(4L_1^2+\lambda^{-3}L_2^2+\lambda^{-3}L_3^2\right)\ddot{\lambda}+\frac{1}{8}\rho\dot{\lambda}^2V\left(L_2^2+L_3^2\right)\lambda^{-4}$$
$$+V\left\{2\left(C_{10}\lambda+C_{01}\right)\left(1-\lambda^{-3}\right)-\frac{1}{2}\varepsilon E_3^2\right\}-2k_1\left(\lambda^2\cos^2\alpha+\lambda^{-1}\sin^2\alpha-1\right)$$
$$\left(2\lambda\cos^2\alpha-\lambda^{-2}\sin^2\alpha\right)\exp\left[k_2\left(\lambda^2\cos^2\alpha+\lambda^{-1}\sin^2\alpha-1\right)^2-1\right]=0 \tag{5.55}$$

$$\frac{1}{12}\left(4L_1^2+\lambda^{-3}L_2^2+\lambda^{-3}L_3^2\right)\ddot{\lambda}+\frac{1}{8}\dot{\lambda}^2\left(L_2^2+L_3^2\right)\lambda^{-4}$$
$$+\left\{2\left(C_{10}\lambda+C_{01}\right)\left(1-\lambda^{-3}\right)-\frac{1}{2}\varepsilon E_3^2\right\}-2k_1\left(\lambda^2\cos^2\alpha+\lambda^{-1}\sin^2\alpha-1\right)$$
$$\left(2\lambda\cos^2\alpha-\lambda^{-2}\sin^2\alpha\right)\exp\left[k_2\left(\lambda^2\cos^2\alpha+\lambda^{-1}\sin^2\alpha-1\right)^2-1\right]=0 \tag{5.56}$$

### 5.4.2 Pure Shear

The same procedure is followed to drive the equation of motion for a DEA under pure shear loading:

$$\boldsymbol{F}=\begin{bmatrix}\lambda & 0 & 0\\ 0 & 1 & 0\\ 0 & 0 & \frac{1}{\sqrt{\lambda}}\end{bmatrix} \tag{5.57}$$

$$C = \begin{bmatrix} \lambda^2 & 0 & 0 \\ 0 & 1 & 0 \\ 0 & 0 & \lambda^{-1} \end{bmatrix} \tag{5.58}$$

$$I_1 = \lambda^2 + 1 + \lambda^{-1} \tag{5.59}$$

$$I_2 = \lambda^2 + \lambda + \lambda^{-1} \tag{5.60}$$

$$C^{-1} = \begin{bmatrix} \lambda^{-2} & 0 & 0 \\ 0 & 1 & 0 \\ 0 & 0 & \lambda \end{bmatrix} \tag{5.61}$$

These expressions, in addition to Equations (5.62) and (5.63), are employed to drive electric and anisotropic terms of free energy.

$$\boldsymbol{C}^{-1} : (\mathbb{E} \otimes \mathbb{E}) = \boldsymbol{\lambda E_3^2} \tag{5.62}$$

$$I_4 = I_6 = \boldsymbol{A}^{(1)} \cdot \boldsymbol{C}\boldsymbol{A}^{(1)} = \lambda^2 \cos^2 \alpha + \sin^2 \alpha \tag{5.63}$$

Free energy definition leads to the potential energy:

$$U = \int_V W dV = L_1 L_2 L_3 \Big\{ C_{10} \left(\lambda^2 + 1 + \lambda^{-1} - 3\right) + C_{01} \left(\lambda^2 + 1 + \lambda^{-1} - 3\right) - \frac{1}{2} \varepsilon \lambda E_3^2$$
$$+ \frac{k_1}{k_2} \left( \exp\left[ k_2 \left(\lambda^2 \cos^2 \alpha + \sin^2 \alpha - 1\right)^2 \right] - 1 \right) \Big\} \tag{5.64}$$

According to the defined deformation gradient in (5.57), displacement and velocity along the axis are:

$$x_1 = \lambda(t) X_1$$
$$x_2 = X_2$$
$$x_3 = \lambda^{-\frac{1}{2}} X_3 \tag{5.65}$$

$$\dot{x}_1 = \dot{\lambda} X_1$$
$$\dot{x}_2 = 0$$
$$\dot{x}_3 = -\frac{1}{2} \dot{\lambda} \lambda^{-\frac{3}{2}} X_3 \tag{5.66}$$

Velocity vectors are utilized for the kinetic energy definition:

$$\begin{aligned}\boldsymbol{K} &= \frac{1}{2}\rho\int_0^{L_3}\int_0^{L_2}\int_0^{L_1}\left(\dot{x}_1^2+\dot{x}_3^2\right)dX_1dX_2dX_3 = \frac{1}{2}\rho\int_0^{L_3}\int_0^{L_2}\int_0^{L_1}\dot{x}_1^2dX_1dX_2dX_3\\ &+\frac{1}{2}\rho\int_0^{L_3}\int_0^{L_2}\int_0^{L_1}\dot{x}_3^2dX_1dX_2dX_3 = \frac{1}{2}\rho\dot{\lambda}^2\int_0^{L_3}\int_0^{L_2}\int_0^{L_1}X_1^2dX_1dX_2dX_3\\ &+\frac{1}{8}\rho\dot{\lambda}^2\lambda^{-3}\int_0^{L_3}\int_0^{L_2}\int_0^{L_1}X_3^2dX_1dX_2dX_3 = \frac{1}{6}\rho\dot{\lambda}^2L_1^3L_2L_3\\ &+\frac{1}{24}\rho\dot{\lambda}^2\lambda^{-3}L_3^3L_2L_1 = \frac{1}{24}\rho\dot{\lambda}^2L_1L_2L_3\left(4L_1^2+\lambda^{-3}L_3^2\right)\end{aligned} \tag{5.67}$$

And we have the Lagrangian in (5.68)

$$\begin{aligned}\mathcal{L} = K-U &= \frac{1}{24}\rho\dot{\lambda}^2L_1L_2L_3\left(4L_1^2+\lambda^{-3}L_3^2\right)\\ &-V\left\{\left(C_{10}+C_{01}\right)\left(\lambda^2+\lambda^{-1}-2\right)-\frac{1}{2}\varepsilon\lambda E_3^2\right.\\ &\left.+\frac{k_1}{k_2}\left(\exp\left[k_2\left(\lambda^2\cos^2\alpha+\sin^2\alpha-1\right)^2\right]-1\right)\right\}\end{aligned} \tag{5.68}$$

Additional expressions are: obtained:

$$\begin{aligned}\frac{\partial\mathcal{L}}{\partial\lambda} &= -\frac{1}{8}\rho\dot{\lambda}^2V\lambda^{-4}L_3^2\\ &-V\left\{\left(C_{10}+C_{01}\right)\left(2\lambda-\lambda^{-2}\right)-\frac{1}{2}\varepsilon E_3^2+4k_1\lambda\cos^2\alpha\right.\\ &\left.\left(\lambda^2\cos^2\alpha+\sin^2\alpha-1\right)\left(\exp\left[k_2\left(\lambda^2\cos^2\alpha+\sin^2\alpha-1\right)^2\right]\right)\right\}\end{aligned} \tag{5.69}$$

$$\frac{\partial\mathcal{L}}{\partial\dot{\lambda}} = \frac{1}{12}\rho L_1L_2L_3\left(4L_1^2+\lambda^{-3}L_3^2\right)\dot{\lambda} \tag{5.70}$$

$$\frac{\partial}{\partial t}\left(\frac{\partial\mathcal{L}}{\partial\dot{\lambda}}\right) = \frac{1}{12}\rho L_1L_2L_3\left(4L_1^2+\lambda^{-3}L_3^2\right)\ddot{\lambda} \tag{5.71}$$

Finally, the Euler-Lagrange equation would be:

$$\begin{aligned}&\frac{1}{12}\rho V\left(4L_1^2+\lambda^{-3}L_3^2\right)\ddot{\lambda}+\frac{1}{8}\rho\dot{\lambda}^2V\lambda^{-4}L_3^2\\ &+V\left\{\left(C_{10}+C_{01}\right)\left(2\lambda-\lambda^{-2}\right)-\frac{1}{2}\varepsilon E_3^2+4k_1\lambda\cos^2\alpha\right.\\ &\left.\left(\lambda^2\cos^2\alpha+\sin^2\alpha-1\right)\exp\left[k_2\left(\lambda^2\cos^2\alpha+\sin^2\alpha-1\right)^2\right]\right\} = \mathbf{0}\end{aligned} \tag{5.72}$$

## 5.5 ENERGY BALANCE

The Euler-Lagrange equation presents a weak form of equilibrium. It is a particular expression of system energy variations. Another standard method to extract the equation of motion is to use thermodynamic laws for energy balance in the system. According to thermodynamic laws, free energy changes of a system, e.g., a DEA, equal the work done by external $\Delta F$, internal forces, $\Delta P$, damping forces, $\Delta G$, and applied voltage $\Delta\phi$:

$$V\delta W = \Delta F - \Delta P - \Delta G + \Delta\phi \tag{5.73}$$

They are expressed as:

$$\Delta F = F_1 L_1 \delta\lambda_1 + F_2 L_2 \delta\lambda_2 \tag{5.74}$$

$$\begin{aligned} \Delta P &= \int_0^{L_1} \rho L_2 L_3 x_1^2 \delta\lambda_1 \frac{d^2\lambda_1}{dt^2} dx_1 + \int_0^{L_2} \rho L_1 L_3 x_2^2 \delta\lambda_1 \frac{d^2\lambda_2}{dt^2} dx_2 \\ &= \frac{1}{3}\rho V\left( L_1^2 \frac{d^2\lambda_1}{dt^2}\delta\lambda_1 + L_2^2 \frac{d^2\lambda_2}{dt^2}\delta\lambda_2 \right) \end{aligned} \tag{5.75}$$

$$\Delta G = \int_0^{L_1} c x_1 \delta\lambda_1 \frac{d\lambda_1}{dt} dx_1 + \int_0^{L_2} c x_2 \delta\lambda_2 \frac{d\lambda_2}{dt} dx_2 = \frac{1}{2}c\left( L_1^2 \delta\lambda_1 \frac{d\lambda_1}{dt} + L_2^2 \delta\lambda_2 \frac{d\lambda_2}{dt} \right) \tag{5.76}$$

$$\Delta\phi = \mathbb{V}\delta Q \tag{5.77}$$

where, $F_i$, c and $Q$ are tensile forces, damping coefficient and electric charges, respectively.

The electric charges distributed on the DE film are obtained as follows:

$$Q = \varepsilon E_3 l_1 l_2 = \varepsilon \frac{\mathbb{V}}{l_3} l_1 l_2 = \varepsilon \frac{\mathbb{V}}{\lambda_3 L_3} \lambda_1 L_1 \lambda_2 L_2 = \varepsilon \frac{\mathbb{V}}{L_3} \lambda_1^2 \lambda_2^2 L_1 L_2 \tag{5.78}$$

And then

$$\delta Q = \varepsilon \frac{\delta\mathbb{V}}{L_3} \lambda_1^2 \lambda_2^2 L_1 L_2 + 2\varepsilon \frac{\mathbb{V}}{L_3} \lambda_1 \delta\lambda_1 \lambda_2^2 L_1 L_2 + 2\varepsilon \frac{\mathbb{V}}{L_3} \lambda_2 \delta\lambda_2 \lambda_1^2 L_1 L_2 \tag{5.79}$$

$$\Delta\phi = \varepsilon \frac{\mathbb{V}\delta\mathbb{V}}{L_3} \lambda_1^2 \lambda_2^2 L_1 L_2 + 2\varepsilon \frac{\mathbb{V}^2}{L_3} \lambda_1 \delta\lambda_1 \lambda_2^2 L_1 L_2 + 2\varepsilon \frac{\mathbb{V}^2}{L_3} \lambda_2 \delta\lambda_2 \lambda_1^2 L_1 L_2 \tag{5.80}$$

Furthermore, to develop the free energy function, the Mooney-Rivlin model is replaced by Gent model in previous relations:

$$W = W^{\text{hyper}} + W^{\text{dielectric}} + W^{\text{aniso}} = -\frac{1}{2}\mu J_m \ln\left(1 - \frac{I_1 - 3}{J_m}\right) - \frac{1}{2}\varepsilon J C^{-1} : (\mathbb{E} \otimes \mathbb{E})$$
$$+ \frac{k_1}{2k_2}\sum_{i=4,6}\left\{\exp\left[k_2 (I_i - 1)^2\right] - 1\right\} = -\frac{1}{2}\mu J_m \ln\left(1 - \frac{\lambda_1^2 + \lambda_2^2 + \lambda_1^{-2}\lambda_2^{-2} - 3}{J_m}\right)$$
$$+ \frac{1}{2}\varepsilon\lambda_1^2\lambda_2^2\left(\frac{\mathbb{V}}{L_3}\right)^2 + \frac{k_1}{k_2}\left\{\exp\left[k_2\left(\lambda_1^2\cos^2\alpha + \lambda_2^2\sin^2\alpha - 1\right)^2\right] - 1\right\} \tag{5.81}$$

$$\delta W = \mu\frac{\lambda_1\delta\lambda_1 + \lambda_2\delta\lambda_2 - \lambda_1^{-3}\lambda_2^{-2}\delta\lambda_1 - \lambda_1^{-2}\lambda_2^{-3}\delta\lambda_2}{1 - \dfrac{\lambda_1^2 + \lambda_2^2 + \lambda_1^{-2}\lambda_2^{-2} - 3}{J_m}} + \varepsilon\lambda_1^2\lambda_2^2\left(\frac{1}{L_3}\right)^2 \mathbb{V}\delta\mathbb{V}$$
$$+ \varepsilon\lambda_1\delta\lambda_1\lambda_2^2\left(\frac{\mathbb{V}}{L_3}\right)^2 + \varepsilon\lambda_1^2\lambda_2\delta\lambda_2\left(\frac{\mathbb{V}}{L_3}\right)^2 + 2k_1\left(\lambda_1^2\cos^2\alpha + \lambda_2^2\sin^2\alpha - 1\right)$$
$$\left(2\lambda_1\delta\lambda_1\cos^2\alpha + \lambda_2\delta\lambda_2\sin^2\alpha\right)\left\{\exp\left[k_2\left(\lambda_1^2\cos^2\alpha + \lambda_2^2\sin^2\alpha - 1\right)^2\right] - 1\right\} \tag{5.82}$$

Substituting these expressions in (5.47) leads to the following:

$$V\left(\mu\frac{\lambda_1\delta\lambda_1 + \lambda_2\delta\lambda_2 - \lambda_1^{-3}\lambda_2^{-2}\delta\lambda_1 - \lambda_1^{-2}\lambda_2^{-3}\delta\lambda_2}{1 - \dfrac{\lambda_1^2 + \lambda_2^2 + \lambda_1^{-2}\lambda_2^{-2} - 3}{J_m}} + \varepsilon\lambda_1^2\lambda_2^2\left(\frac{1}{L_3}\right)^2 \mathbb{V}\delta\mathbb{V}\right.$$
$$+ \varepsilon\lambda_1\delta\lambda_1\lambda_2^2\left(\frac{\mathbb{V}}{L_3}\right)^2 + \varepsilon\lambda_1^2\lambda_2\delta\lambda_2\left(\frac{\mathbb{V}}{L_3}\right)^2$$
$$+ 2k_1\left(\lambda_1^2\cos^2\alpha + \lambda_2^2\sin^2\alpha - 1\right)\left(2\lambda_1\delta\lambda_1\cos^2\alpha + \lambda_2\delta\lambda_2\sin^2\alpha\right)$$
$$\left.\left\{\exp\left[k_2\left(\lambda_1^2\cos^2\alpha + \lambda_2^2\sin^2\alpha - 1\right)^2\right] - 1\right\}\right) = P_1 L_1\delta\lambda_1 + P_2 L_2\delta\lambda_2$$
$$-\frac{1}{3}\rho V\left(L_1^2\frac{d^2\lambda_1}{dt^2}\delta\lambda_1 + L_2^2\frac{d^2\lambda_2}{dt^2}\delta\lambda_2\right) - \frac{1}{2}c\left(L_1^2\delta\lambda_1\frac{d\lambda_1}{dt} + L_1^2\delta\lambda_1\frac{d\lambda_1}{dt}\right)$$
$$+ \varepsilon\frac{\mathbb{V}\delta\mathbb{V}}{L_3}\lambda_1^2\lambda_2^2 L_1 L_2 + 2\varepsilon\frac{\mathbb{V}^2}{L_3}\lambda_1\delta\lambda_1\lambda_2^2 L_1 L_2 + 2\varepsilon\frac{\mathbb{V}^2}{L_3}\lambda_2\delta\lambda_2\lambda_1^2 L_1 L_2 \tag{5.83}$$

$$\mu\frac{\lambda_1\delta\lambda_1+\lambda_2\delta\lambda_2-\lambda_1^{-3}\lambda_2^{-2}\delta\lambda_1-\lambda_1^{-2}\lambda_2^{-3}\delta\lambda_2}{1-\dfrac{\lambda_1^2+\lambda_2^2+\lambda_1^{-2}\lambda_2^{-2}-3}{J_m}}+2k_1\left(\lambda_1^2\cos^2\alpha+\lambda_2^2\sin^2\alpha-1\right)$$
$$\left(2\lambda_1\delta\lambda_1\cos^2\alpha+\lambda_2\delta\lambda_2\sin^2\alpha\right)\left\{\exp\left[k_2\left(\lambda_1^2\cos^2\alpha+\lambda_2^2\sin^2\alpha-1\right)^2\right]-1\right\}$$
$$=\sigma_1\delta\lambda_1+\sigma_2\delta\lambda_2-\frac{1}{3}\rho\left(L_1^2\frac{d^2\lambda_1}{dt^2}\delta\lambda_1+L_2^2\frac{d^2\lambda_2}{dt^2}\delta\lambda_2\right)$$
$$-\frac{1}{2}c\left(\frac{L_1}{L_2L_3}L_1^2\delta\lambda_1\frac{d\lambda_1}{dt}+\frac{L_2}{L_1L_3}\delta\lambda_2\frac{d\lambda_2}{dt}\right)+\varepsilon\frac{\mathbb{V}^2}{L_3^2}\lambda_1\delta\lambda_1\lambda_2^2+\varepsilon\frac{\mathbb{V}^2}{L_3^2}\lambda_2\delta\lambda_2\lambda_1^2 \tag{5.84}$$

$$\frac{1}{3}\rho\left(L_1^2\frac{d^2\lambda_1}{dt^2}\right)+\frac{1}{2}c\left(\frac{L_1}{L_2L_3}L_1^2\frac{d\lambda_1}{dt}\right)-\varepsilon\frac{\mathbb{V}^2}{L_3^2}\lambda_1\lambda_2^2$$
$$+\mu\frac{\lambda_1-\lambda_1^{-3}\lambda_2^{-2}}{1-\dfrac{\lambda_1^2+\lambda_2^2+\lambda_1^{-2}\lambda_2^{-2}-3}{J_m}}+4\cos^2\alpha k_1\lambda_1\left(\lambda_1^2\cos^2\alpha+\lambda_2^2\sin^2\alpha-1\right)$$
$$\left\{\exp\left[k_2\left(\lambda_1^2\cos^2\alpha+\lambda_2^2\sin^2\alpha-1\right)^2\right]-1\right\}=\sigma_1 \tag{5.85}$$

$$\frac{1}{3}\rho\left(L_2^2\frac{d^2\lambda_2}{dt^2}\right)-\frac{1}{2}c\left(\frac{L_2}{L_1L_3}\frac{d\lambda_2}{dt}\right)-\varepsilon\frac{\mathbb{V}^2}{L_3^2}\lambda_2\lambda_1^2$$
$$+\mu\frac{\lambda_2-\lambda_1^{-2}\lambda_2^{-3}}{1-\dfrac{\lambda_1^2+\lambda_2^2+\lambda_1^{-2}\lambda_2^{-2}-3}{J_m}}+4\sin^2\alpha k_1\lambda_2\left(\lambda_1^2\cos^2\alpha+\lambda_2^2\sin^2\alpha-1\right)$$
$$\left\{\exp\left[k_2\left(\lambda_1^2\cos^2\alpha+\lambda_2^2\sin^2\alpha-1\right)^2\right]-1\right\}=\sigma_2 \tag{5.86}$$

## 5.6 VIRTUAL WORK PRINCIPLE

Same expression could be obtained using virtual work principle: we assume virtual displacements, $\delta u_1 = L_1\delta\lambda_1$ and $\delta u_2 = L_2\delta\lambda_2$ and virtual electric displacement to be $\delta \mathbb{d}$. $\mathbb{d}$, the electric displacement in current is defined as:

$$\mathbb{d}=\frac{Q}{l_1l_2}=\frac{Q}{\lambda_1L_1\lambda_2L_2} \tag{5.87}$$

And then virtual charges are:

$$\delta Q=\lambda_1L_1\lambda_2L_2\delta\mathbb{d}+\delta\lambda_1L_1\lambda_2L_2\mathbb{d}+\lambda_1L_1\delta\lambda_2L_2\mathbb{d} \tag{5.88}$$

Then, the virtual external work done because of displacements by mechanical loads, $F_1$ and $F_2$, and voltage, $\mathbb{V}$ are:

$$\delta E = F_1\delta u_1 - F_1(-\delta u_1) + F_2\delta u_2 - F_2(-\delta u_2) + \mathbb{V}\delta Q = 2F_1\delta u_1 + 2F_2\delta u_2 + \mathbb{V}(\lambda_1 L_1\lambda_2 L_2\delta \mathrm{d} + \delta\lambda_1 L_1\lambda_2 L_2\mathrm{d} + \lambda_1 L_1\delta\lambda_2 L_2\mathrm{d}) \tag{5.89}$$

On the other hand, due to the deformation, internal virtual work is done:

$$\delta I = \int \delta P_{ij} S_{ij} dV = \delta P_{ij} S_{ij} V \tag{5.90}$$

where $\boldsymbol{S}$ is the nominal stress tensor. For a homogeneous bi-axial loading, $\delta I$ is expressed as:

$$\delta I = (\delta P_{ij} S_{ij} + \delta P_{ij} S_{ij})V = (\delta\lambda_1 S_{11} + \delta\lambda_2 S_{22})V \tag{5.91}$$

The kinetic energy changes according to the following:

$$\begin{aligned}\delta K &= \rho\int_0^{L_3}\int_0^{L_2}\int_0^{L_1}\ddot{x}_1\delta x_1 dX_1 dX_2 dX_3 + \rho\int_0^{L_3}\int_0^{L_2}\int_0^{L_1}\ddot{x}_2\delta x_2 dX_1 dX_2 dX_3 \\ &+\rho\int_0^{L_3}\int_0^{L_2}\int_0^{L_1}\ddot{x}_3\delta x_3 dX_1 dX_2 dX_3 = \rho\int_0^{L_3}\int_0^{L_2}\int_0^{L_1}\ddot{\lambda}_1 X_1\delta\lambda_1 X_1 dX_1 dX_2 dX_3 \\ &+\rho\int_0^{L_3}\int_0^{L_2}\int_0^{L_1}\ddot{\lambda}_2 X_2\delta\lambda_2 X_2 dX_1 dX_2 dX_3 = \rho\ddot{\lambda}_1\delta\lambda_1\int_0^{L_3}\int_0^{L_2}\int_0^{L_1} X_1^2 dX_1 dX_2 dX_3 \\ &+\rho\ddot{\lambda}_2\delta\lambda_2\int_0^{L_3}\int_0^{L_2}\int_0^{L_1} X_2^2 dX_1 dX_2 dX_3 = \frac{1}{3}\rho\left(\ddot{\lambda}_1\delta\lambda_1 L_1^3 L_2 L_3 + \ddot{\lambda}_2\delta\lambda_2 L_2^3 L_1 L_3\right)\end{aligned} \tag{5.92}$$

For the equilibrium state, we have:

$$\delta I - \delta E + \delta K = 0 \tag{5.93}$$

$$\begin{aligned}&(\delta\lambda_1 S_{11} + \delta\lambda_2 S_{22})V - 2F_1 L_1\delta\lambda_1 + 2F_2 L_2\delta\lambda_2 \\ &+\mathbb{V}(\lambda_1 L_1\lambda_2 L_2\delta \mathrm{d} + \delta\lambda_1 L_1\lambda_2 L_2\mathrm{d} + \lambda_1 L_1\delta\lambda_2 L_2\mathrm{d}) \\ &+\frac{1}{3}\rho\left(\ddot{\lambda}_1\delta\lambda_1 L_1^3 L_2 L_3 + \ddot{\lambda}_2\delta\lambda_2 L_2^3 L_1 L_3\right) = 0\end{aligned} \tag{5.94}$$

Final governing equations are obtained regarding the independent variations:

$$\frac{1}{3}\rho\ddot{\lambda}_1 L_1^2 = \frac{2}{L_2 L_3}F_1 - \frac{1}{L_1}\mathbb{V}\lambda_2\mathrm{d} - S_{11} \tag{5.95}$$

$$\frac{1}{3}\rho\ddot{\lambda}_2 L_2^2 = \frac{2}{L_1 L_3}F_2 - \frac{1}{L_2}\mathbb{V}\lambda_1\mathrm{d} - S_{22} \tag{5.96}$$

These equations include the stress components which are functions of stretch. Derived constitutive law in Chapter 3 are used to describe the stress and stretch relation. Then the equations could be solved.

## 5.7 CONCLUSION

Once the Helmholtz free energy of a dielectric elastomer is defined, it could be used to determine the equation of motion. Therefore, this chapter presents the governing equations for the dynamic and transient analysis of dielectric elastomer actuators. Initially, the main parameters describing the state of the actuator are determined. Then, the governing equations, considering the whole system, including the electric excitation and external mechanical forces, are derived using the Euler-Lagrange equation with and without rate dependency. More detailed equations are: presented for two cases of uniaxial tensile and pure shear loading, as well. The equations derivation for a general case using energy balance and virtual work principle are also included.

## REFERENCES

[1] J. Zhu, S. Cai, Z. Suo, Nonlinear oscillation of a dielectric elastomer balloon, *Polymer International*, 59(3) (2010) 378–383.

[2] T. Li, S. Qu, W. Yang, Electromechanical and dynamic analyses of tunable dielectric elastomer resonator, *International Journal of Solids and Structures*, 49(26) (2012) 3754–3761.

[3] P. Dubois, S. Rosset, M. Niklaus, M. Dadras, H. Shea, Voltage control of the resonance frequency of dielectric electroactive polymer (DEAP) membranes, *Journal of Microelectromechanical Systems*, 17(5) (2008) 1072–1081.

[4] Y. Wang, C. Feng, J. Yang, D. Zhou, S. Wang, Nonlinear vibration of FG-GPLRC dielectric plate with active tuning using differential quadrature method, *Computer Methods in Applied Mechanics and Engineering*, 379 (2021) 113761.

[5] J. Zhu, S. Cai, Z. Suo, Resonant behavior of a membrane of a dielectric elastomer, *International Journal of Solids and Structures*, 47(24) (2010) 3254–3262.

[6] E. Rustighi, W. Kaal, S. Herold, A. Kubbara, Experimental characterisation of a flat dielectric elastomer loudspeaker, *Actuators*, 7 2018 28.

[7] N. Hosoya, H. Masuda, S. Maeda, Balloon dielectric elastomer actuator speaker, *Applied Acoustics*, 148 (2019) 238–245.

[8] E. Garnell, O. Doaré, C. Rouby, Coupled vibro-acoustic modeling of a dielectric elastomer loudspeaker, *The Journal of the Acoustical Society of America*, 147(3) (2020) 1812–1821.

[9] P. Linnebach, G. Rizzello, S. Seelecke, Design and validation of a dielectric elastomer membrane actuator driven pneumatic pump, *Smart Materials and Structures*, 29(7) (2020) 075021.

[10] C. Cao, X. Gao, A.T. Conn, A magnetically coupled dielectric elastomer pump for soft robotics, *Advanced Materials Technologies*, 4(8) (2019) 1900128.

[11] S. Ho, H. Banerjee, Y.Y. Foo, H. Godaba, W.M.M. Aye, J. Zhu, C.H. Yap, Experimental characterization of a dielectric elastomer fluid pump and optimizing performance via composite materials, *Journal of Intelligent Material Systems and Structures*, 28(20) (2017) 3054–3065.

[12] B. Li, J. Zhang, L. Liu, H. Chen, S. Jia, D. Li, Modeling of dielectric elastomer as electromechanical resonator, *Journal of Applied Physics*, 116(12) (2014).

[13] C. Cao, S. Burgess, A.T. Conn, Toward a dielectric elastomer resonator driven flapping wing micro air vehicle, *Frontiers in Robotics and AI*, 5 (2019) 137.

[14] C. Feng, L. Yu, W. Zhang, Dynamic analysis of a dielectric elastomer-based microbeam resonator with large vibration amplitude, *International Journal of Non-Linear Mechanics*, 65 (2014) 63–68.

[15] J. Fox, N. Goulbourne, On the dynamic electromechanical loading of dielectric elastomer membranes, *Journal of the Mechanics and Physics of Solids*, 56(8) (2008) 2669–2686.

[16] G. Moretti, G. Rizzello, M. Fontana, S. Seelecke, High-frequency voltage-driven vibrations in dielectric elastomer membranes, *Mechanical Systems and Signal Processing*, 168 (2022) 108677.

[17] A. Tsujino, T. Hayakawa, Analysis of Dynamic Characteristics of a Balloon-Type Dielectric Elastomer Actuator Pre-Stretched by Water Pressure, in: *2023 IEEE/SICE International Symposium on System Integration (SII)*, IEEE, 2023, pp. 1–4.

[18] C. Feng, L. Jiang, W.M. Lau, Dynamic characteristics of a dielectric elastomer-based microbeam resonator with small vibration amplitude, *Journal of Micromechanics and Microengineering*, 21(9) (2011) 095002.

[19] A. Alibakhshi, H. Heidari, Nonlinear dynamic responses of electrically actuated dielectric elastomer-based microbeam resonators, *Journal of Intelligent Material Systems and Structures*, 33(4) (2022) 558–571.

[20] D. Kumar, S. Sarangi, Dynamic modeling of a dielectric elastomeric spherical actuator: An energy-based approach, *Soft Materials*, 19(2) (2021) 129–138.

[21] H. Yong, X. He, Y. Zhou, Dynamics of a thick-walled dielectric elastomer spherical shell, *International Journal of Engineering Science*, 49(8) (2011) 792–800.

[22] Y. Li, I. Oh, J. Chen, H. Zhang, Y. Hu, Nonlinear dynamic analysis and active control of visco-hyperelastic dielectric elastomer membrane, *International Journal of Solids and Structures*, 152 (2018) 28–38.

[23] J. Zhang, H. Chen, D. Li, Nonlinear dynamical model of a soft viscoelastic dielectric elastomer, *Physical Review Applied*, 8(6) (2017) 064016.

[24] J. Zhang, H. Chen, B. Li, D. McCoul, Q. Pei, Coupled nonlinear oscillation and stability evolution of viscoelastic dielectric elastomers, *Soft Matter*, 11(38) (2015) 7483–7493.

[25] M. Tewary, T. Roy, Nonlinear dynamic analysis of anisotropic bimorph dielectric elastomer actuator for soft fish robots, *Communications in Nonlinear Science and Numerical Simulation*, 127 (2023) 107585.

[26] K. Kashyap, A.K. Sharma, M.M. Joglekar, Nonlinear dynamic analysis of aniso-visco-hyperelastic dielectric elastomer actuators, *Smart Materials and Structures*, 29(5) (2020) 055014.

[27] B.-X. Xu, R. Mueller, A. Theis, M. Klassen, D. Gross, Dynamic analysis of dielectric elastomer actuators, *Applied Physics Letters*, 100(11) (2012) 935–938.

[28] J. Sheng, H. Chen, L. Liu, J. Zhang, Y. Wang, S. Jia, Dynamic electromechanical performance of viscoelastic dielectric elastomers, *Journal of Applied Physics*, 114(13) (2013) 134101.

# 6 Computational Aspects

## 6.1 INTRODUCTION

Previous chapters reviewed fundamental ideas to study the behavior of a dielectric elastomer. We introduced the framework required to model different aspects of the material behavior in Chapters 5 and 6. Then, in Chapter 7, the equation of motion for such a material in a system was devised. The current chapter is dedicated to the analysis procedure using finite element methods.

As a common tool, Abaqus simulates a mechanical problem including the 3D geometrical body, material properties, boundary condition and loadings and discretizing. This steps shape pre-process phase. In the next phase, the software solves equilibrium equations for the discretized body using iterative procedures. It finally presents requested results as standard data base or visualized presentation. Provided information in the post- process phase let the user to interpret the results.

In addition to built-in features, Abaqus let the user to strength it by some user-defined subroutines. The subroutines are powerful tools to design customized features. Therefore, this chapter mainly discuses numerical methods and employed subroutine essentials employed by finite Abaqus.

## 6.2 ABAQUS USER SUBROUTINES, UMAT

Despite the vast and various built in features in Abaqus, in some simulations the user needs to develop more material models, elements, load or properties. In these cases, subroutines allow the user to customize software functionality for the specific desire. All the subroutines are written and compiled in Fortran. They also have particular interface that is essential to obeyed for proper implementation.

In addition to some specific variables the user has to define, some subroutines allow the user to define state variables, `STATEV`. State variables are functions of other variables in the subroutine.

To model the behavior of materials using the constitutive law and analyzing the problem by Abaqus/Standard, UMAT subroutines are utilized. Once a user-defined material subroutine is included for an element, it is called for the material calculation points of the element. At the end of each increment, the subroutine updates stress, material Jacobian matrix and all the solution-dependent state variables.


DOI: 10.1201/9781003571469-8

A UMAT must include DDSDDE, Jacobian matrix of the constitutive model and STRESS, Cauchy stress according to the user manual. The user manual also presents the standard interface of the UMAT subroutine as follows:

```
SUBROUTINE UMAT(STRESS,STATEV,DDSDDE,SSE,SPD,SCD,
      1 RPL,DDSDDT,DRPLDE,DRPLDT,
      2 STRAN,DSTRAN,TIME,DTIME,TEMP,DTEMP,PREDEF,
        DPRED,CMNAME,
      3 NDI,NSHR,NTENS,NSTATV,PROPS,NPROPS,COORDS,
        DROT,PNEWDT,
      4 CELENT,DFGRD0,DFGRD1,NOEL,NPT,LAYER,KSPT,
        KSTEP,KINC)
C
       INCLUDE 'ABA_PARAM.INC'
C
       CHARACTER*80 CMNAME
       DIMENSION STRESS(NTENS),STATEV(NSTATV),
      1 DDSDDE(NTENS,NTENS),DDSDDT(NTENS),DRPLDE
        (NTENS),
      2 STRAN(NTENS),DSTRAN(NTENS),TIME
        (2),PREDEF(1),
        DPRED(1),
      3 PROPS(NPROPS),COORDS(3),DROT(3,3),DFGRD0
        (3,3),DFGRD1(3,3)

        user coding to define DDSDDE, STRESS,
         STATEV, SSE, SPD, SCD
        and, if necessary, RPL, DDSDDT, DRPLDE,
         DRPLDT, PNEWDT

        RETURN
        END
```

First part of the interface represents an argument list to pass the variables to the user subroutine. The argument list is followed by the statement: "INCLUDE 'ABA_PARAM.INC'". This statement calls ABA_PARAM.INC' file which contains installation parameters and compiles the subroutine and other tools of the software.

The subsequent line determines the material user-specified name via "CMNAME" followed by lines stablishing the dimension for each variable.

Then, there is the main part of the subroutine where the user defines essential variables. This part usually starts with the introducing the model constants as "PROPS". Some users prefer to define frequently-used constants, as well. Then it is followed by the definition of "STRESS", "DDSDDE", and other required variables. Section 6.5 UMAT Implementation reviews this part for an example constitutive law.

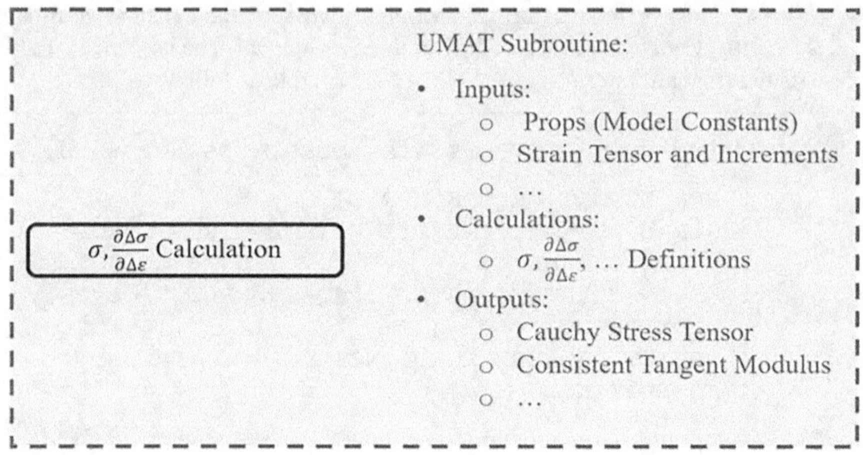


**FIGURE 6.1** UMAT details.

At the end, the phrases "`RETURN`" and "`RETURN`" make the Abaqus/Standard to read the UMAT as many times as needed and stop it finally. Figure 6.1 briefly depicts how a UMAT performs.

variables in the argument list and other terms used in UMATS [1].

Table 6.1 summarizes used variables in the argument list and other terms used in UMATS [1].

## TABLE 6.1
## UMAT Variables and Terms

| # | Variable | Definition |
|---|---|---|
| 1 | STRESS | "True" (Cauchy) Stress |
| 2 | STATEV | Solution-dependent State Variables |
| 3 | DDSDDE | Jacobian Matrix of the Constitutive Model |
| 4 | SSE | Specific Elastic Strain Energy |
| 5 | SPD | Specific Plastic Dissipation Energy |
| 6 | SCD | Specific "Creep" Dissipation Energy |
| 7 | RPL | Volumetric Heat Generation per Unit Time |
| 8 | DDSDDT | Variation of the Stress Increments with Respect to the Temperature |
| 9 | DRPLDE | Variation of RPL with Respect to the Strain Increments |
| 10 | DRPLDT | Variation of RPL with Respect to the Temperature |
| 11 | STRAN | Total Strains |
| 12 | DSTRAN | Strain Increments |
| 13 | TIME | Step Time and Total Time |
| 14 | DTIME | Time Increment. |
| 15 | TEMP | Temperature |

(*Continued*)

**TABLE 6.1 (CONTINUED)**
**UMAT Variables and Terms**

| # | Variable | Definition |
|---|---|---|
| 16 | DTEMP | Increment of Temperature |
| 17 | DPRED | Interpolated Values of Predefined Field Variables |
| 18 | PREDEF | Increments of Predefined Field Variables |
| 19 | CMNAME | User-Defined Material Name |
| 20 | NDI | Number of Direct Stress Components |
| 21 | NSHR | Number of Engineering Shear Stress Components |
| 22 | NTENS | Size of the Stress or Strain Component Array |
| 23 | NSTATV | Number of Solution-Dependent State Variables |
| 24 | PROPS | User-Specified Array of Material Constants |
| 25 | NPROPS | User-Defined Number of Material Constants |
| 26 | COORDS | Coordinates |
| 27 | DROT | Rotation Increment Matrix |
| 28 | PNEWDT | Ratio of Suggested New Time Increment to the Time Increment Being Used |
| 29 | CELENT | Characteristic Element Length |
| 30 | DFGRD | Deformation Gradient |
| 31 | NOEL | Element Number |
| 32 | NPT | Integration Point Number |
| 33 | LAYER | Layer Number |
| 34 | KSPT | Section Point Number |
| 35 | KSTEP | Step Number |
| 36 | KINC | Increment Number |

Once the UMAT is developed and introduced to the simulation, Abaqus/Standard analyzes the material's behavior. Figure 6.2 and explains how a UMAT subroutines is employed to analyze a simulation problem.

The following sections are dedicated to the derivation of Cauchy stress and consistent Jacobian matrices and how Abaqus uses them.

## 6.3 CAUCHY (TRUE) STRESS TENSOR

In previous chapters, we reviewed the second Piola-Kirchhoff and Cauchy stress tensors as follows:

$$\boldsymbol{S} = 2\frac{\partial W}{\partial \boldsymbol{C}} \tag{6.1}$$

$$\boldsymbol{\sigma} = \frac{2}{J}\boldsymbol{F}\frac{\partial W}{\partial \boldsymbol{C}}\boldsymbol{F}^T \tag{6.2}$$

Cauchy (or true) stress, $\boldsymbol{\sigma}$ and the second Piola-Kirchhoff stress, $\boldsymbol{S}$ quantify force per unit area in deformed and undeformed configuration, respectively. These stresses

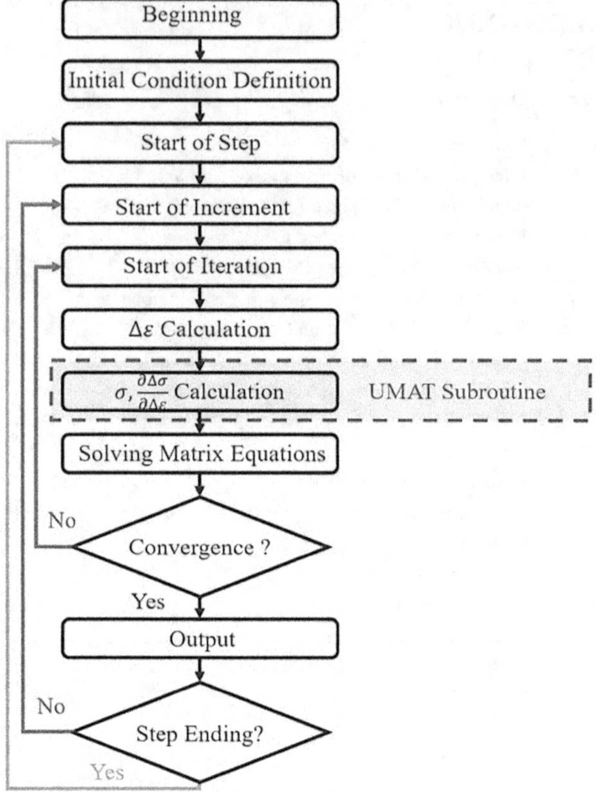


**FIGURE 6.2** Abaqus/standard analysis procedure.

are expressed based on W, Helmholtz free energy function, which is a continuous function of kinematic variables, as mentioned in Chapter 3.

The following equation defines the internal virtual work done by the Cauchy stress tensor:

$$\delta W = \int_V \boldsymbol{\sigma} : \delta \boldsymbol{D} dV \tag{6.3}$$

where $\boldsymbol{D} = \frac{1}{2}\left(\boldsymbol{L} + \boldsymbol{L}^T\right)$, (Equation 4.5) quantifies the rate of deformation and is defined as the symmetric part of $\boldsymbol{L}$, Equation (4.2). The virtual work is defined with respect to the current deformed volume and presents the work per unit deformed volume. In this definition, $\boldsymbol{\sigma}$ and $\boldsymbol{D}$ are said to be work conjugate with respect to the current deformed volume [2].

The need for work per unit volume in different configurations leads to different work conjugate pairs and stress and strain measures. First, we transform the integral in reference configuration and then name the created stress tensor, $J\boldsymbol{\sigma}$, the Kirchhoff stress, $\boldsymbol{\tau}$:

$$\delta W = \int_{V_0} J\boldsymbol{\sigma}: \delta\boldsymbol{D} dV_0 = \int_{V_0} \boldsymbol{\tau}: \delta\boldsymbol{D} dV_0 \quad J\boldsymbol{\sigma} = \boldsymbol{\tau} \tag{6.4}$$

According to the equation, $\boldsymbol{\tau}$ and $\boldsymbol{D}$ are work conjugate pairs with respect to the reference volume.

Replacing $\boldsymbol{D}$ in the integral with $\boldsymbol{L}$, and its equal based on $\boldsymbol{F}$ and $\dot{\boldsymbol{F}}$ leads to

$$\begin{aligned} \delta W &= \int_{V_0} J\boldsymbol{\sigma}: \delta\boldsymbol{L} dV_0 = \int_{V_0} J\boldsymbol{\sigma}: \delta\dot{\boldsymbol{F}}\boldsymbol{F}^{-1} dV_0 = \int_{V_0} J\boldsymbol{\sigma}\boldsymbol{F}^{-T}: \delta\dot{\boldsymbol{F}} dV_0 \\ &= \int_{V_0} \boldsymbol{P}\delta\dot{\boldsymbol{F}} dV_0 \quad J\boldsymbol{\sigma}\boldsymbol{F}^{-T} = \boldsymbol{P} \end{aligned} \tag{6.5}$$

Stress tensor $\boldsymbol{P}$, is named the First Piola- Kirchhoff stress tensor. It is an unsymmetrical two points tensor which means it relates force vector in the deformed configuration to the area vector in the refence configuration (deformed force, per unit of reference area). $\boldsymbol{P}$ and $\dot{\boldsymbol{F}}$ are work conjugates.

Defining Green strain tensor $\boldsymbol{E} = \frac{1}{2}(\boldsymbol{C} - \boldsymbol{I})$ and its rate as:

$$\dot{\boldsymbol{E}} = \frac{1}{2}\dot{\boldsymbol{C}} = \frac{1}{2}\left(\dot{\boldsymbol{F}}^T\boldsymbol{F} + \boldsymbol{F}^T\dot{\boldsymbol{F}}\right) = \frac{1}{2}\left(\boldsymbol{F}^T\boldsymbol{L}^T\boldsymbol{F} + \boldsymbol{F}^T\boldsymbol{L}\boldsymbol{F}\right) = \boldsymbol{F}^T\boldsymbol{D}\boldsymbol{F} \tag{6.6}$$

We have:

$$\boldsymbol{D} = \boldsymbol{F}^{-T}\dot{\boldsymbol{E}}\boldsymbol{F}^{-1} \tag{6.7}$$

Using this relation in Equation (6.4) leads to

$$\begin{aligned} \delta W &= \int_{V_0} J\boldsymbol{\sigma}: \delta\boldsymbol{D} dV_0 = \int_{V_0} J\boldsymbol{\sigma}: \boldsymbol{F}^{-T}\delta\dot{\boldsymbol{E}}\boldsymbol{F}^{-1} dV_0 = \int_{V_0} J\boldsymbol{F}^{-T}\boldsymbol{\sigma}\boldsymbol{F}^{-1}: \delta\dot{\boldsymbol{E}} dV_0 \\ &= \int_{V_0} \boldsymbol{S}: \delta\dot{\boldsymbol{E}} dV_0, \boldsymbol{S} = J\boldsymbol{F}^{-T}\boldsymbol{\sigma}\boldsymbol{F}^{-1} \end{aligned} \tag{6.8}$$

Equation (6.8) reveals the work conjugate of $\boldsymbol{S}$, the second Piola- Kirchhoff stress tensor, which is $\dot{\boldsymbol{E}}$.

## 6.4 CONSISTENT TANGENT MODULUS

Nonlinear behaviors of soft materials, and structures are often investigated numerically and numerical analysis of nonlinear equations includes iterative procedures. Therefore, iterative operative of mechanical parameters are required; e.g. a common technique to solve the nonlinear initial boundary value problems of finite inelasticity is iterative solution of Newton-Raphson method. Algorithmic or consistent tangent modulus is the iteration operator needed for the technique. The "consistent" word in the phrase comes from the concept that the parameter is consistent with the stress update algorithms and demonstrates how algorithmic expression of stresses are sensitive to the variations of total deformation [3].

The consistent tangent modulus is the operative counterpart for the elasticity tensor to quantify the gradient of nonlinear stress with respect to tensorial deformations [3]. The definition of the tangent modulus makes it possible to develop and validate new constitutive models using finite element algorithms. For instance, explicit definition of tangent modulus tensor, in addition to Cauchy stress definition, is essential to implement constitutive models in commercial FE softwares Abaqus and Ansys. Here, we elaborate on the mechanical interpretation of tangent modulus and how it contributes to the iterative algorithms:

The weak form of the virtual work principle for a hyperelastic material, as a deformable solid, is expressed as follows [4]:

$$\int_{V_0} P_{iI} \frac{\partial \delta v_i}{\partial X_I} dV_0 = \int_{V_0} f_i.\delta v_i dV_0 + \int_{\partial V_0} t_i.\delta v_i dS_0 \tag{6.9}$$

$\boldsymbol{P}$, $\boldsymbol{f}$, t and $\delta v$ refer to the first Piola- Kirchhoff stress tensor, body force, surface traction and virtual velocity field, respectively. $V_0$ and $S_0$ denote the volume and surface of the solid.

The left-hand side of the equation is then rewritten using the Kirchhoff stress tensor, $\boldsymbol{\tau}$ and virtual rate of deformation, $\delta \boldsymbol{D}$:

$$\int_{V_0} P_{iI} : \frac{\partial \delta v_i}{\partial X_I} dV_0 = \int_{V_0} P_{iI} : \frac{\partial \delta v_i}{\partial x_k} \frac{\partial x_k}{\partial X_I} dV_0 = \int_{V_0} \tau_{ik} \frac{\partial \delta v_i}{\partial x_k} dV_0 = \int_{V_0} \tau_{ik} \delta D_{iK} dV_0 \tag{6.10}$$

This transformation utilizes the following properties:

$$P_{iI} = \tau_{ik} \frac{\partial X_I}{\partial x_k} \tag{6.11}$$

$$\frac{\partial \delta v_i}{\partial x_k} + \frac{\partial \delta v_k}{\partial x_i} = 2\delta D_{iK} \tag{6.12}$$

$$\frac{\partial \delta v_i}{\partial X_I} = \frac{\partial \delta v_i}{\partial x_k} \frac{\partial x_k}{\partial X_I} = \delta D_{iK} \frac{\partial x_k}{\partial X_I} \tag{6.13}$$

Substituting the new expression of the left-hand side in Equation (6.10), we have [5]:

$$\int_{V_0} \tau_{ik} \delta D_{iK} dV_0 = \int_{V_0} f_i \delta v_i dV_0 + \int_{\partial V_0} t_i \delta v_i dS_0 \tag{6.14}$$

To have a FEM compatible version of the above equation, we consider shape functions, $\tilde{\boldsymbol{N}}$, nodal displacements and nodal virtual velocity field, $u^e$ and $\delta v^e$. To discretize the virtual rate of deformation, we assume $\tilde{\boldsymbol{N}}'$ to be the derivative of $\tilde{\boldsymbol{N}}$ with respect to the deformed configuration:

$$u = \tilde{\boldsymbol{N}} u^e$$

$$\delta v = \tilde{\boldsymbol{N}} \delta v^e$$

$$\delta \boldsymbol{D} = \tilde{\boldsymbol{N}}' \delta v^e \tag{6.15}$$

Presented relations are used to discrete Equation (6.14):

$$\delta v^e \int_{V_0} \boldsymbol{\tau} : \tilde{\boldsymbol{N}}' dV_0 = \delta v^e \int_{V_0} \boldsymbol{f} . \tilde{\boldsymbol{N}} dV_0 + \delta v^e \int_{\partial V_0} \boldsymbol{t} . \tilde{\boldsymbol{N}} dS_0 \tag{6.16}$$

Nodal displacement will be obtained by solving this nonlinear system. An accepted algorithm to solve the system is Newton algorithm. The algorithm calculates displacement $u^k$, at increment $k$ and solve the equation for it. It evaluates the solution and the determined tolerance and, if necessary, computes the new displacement using the relation:

$$u^{k+1} = u^k + \delta u \tag{6.17}$$

The increment, $\delta u$, is defined using the left-hand side of Equation (6.14), named $R$:

$$R = \int_{V_0} \tau_{ik} \delta D_{iK} dV_0 = \int_{V_0} P_{iI} : \frac{\partial \delta v_i}{\partial X_I} dV_0 \tag{6.18}$$

$$\delta u = -\left( \frac{\partial R}{\partial u} \right)^{-1} R\left(u^k\right) = -\boldsymbol{K}^{-1} R\left(u^k\right) \tag{6.19}$$

$\boldsymbol{K}$ is the Jacobian matrix. To this aim, we use the directional variation of $R$ with respect to $\delta u$. Therefore, $u_\epsilon = u + \epsilon \delta u$ leads to:

$$\frac{\partial}{\partial u} R[u] = \left. \frac{d}{d\epsilon} \right|_{\epsilon \to 0} R[u_\epsilon] = \left. \frac{d}{d\epsilon} \right|_{\epsilon \to 0} R[u + \epsilon \delta u] \tag{6.20}$$

Substituting the expression for $R$ from Equations (6.20) in (6.18), we have:

$$\left. \frac{d}{d\epsilon} \int_{V_0} P_{iI} \frac{\partial \delta v_i}{\partial X_I} dV_0 \right|_{\epsilon \to 0} = \left. \int_{V_0} \left( \frac{d}{d\epsilon} P_{iI} \right) \frac{\partial \delta v_i}{\partial X_I} dV_0 \right|_{\epsilon \to 0} = \left. \int_{V_0} \left( \frac{d}{d\epsilon} F_{iM} S_{MI} \right) \frac{\partial \delta v_i}{\partial X_I} dV_0 \right|_{\epsilon \to 0}$$

$$= \left. \int_{V_0} \left( \frac{d}{d\epsilon} F_{iM} \right) S_{MI} \frac{\partial \delta v_i}{\partial X_I} dV_0 \right|_{\epsilon \to 0} \left. \int_{V_0} F_{iM} \left( \frac{d}{d\epsilon} S_{MI} \right) \frac{\partial \delta v_i}{\partial X_I} dV_0 \right|_{\epsilon \to 0} \tag{6.21}$$

Required expressions are:

$$\frac{d}{d\epsilon}F_{iM}\bigg|_{\epsilon\to 0} = \frac{d}{d\epsilon}\frac{\partial x_i}{\partial X_M}\bigg|_{\epsilon\to 0} = \frac{d}{d\epsilon}\frac{\partial\left(X_i + u_i + \epsilon\delta u_i\right)}{\partial X_M}\bigg|_{\epsilon\to 0} = \frac{\partial \delta u_i}{\partial X_M}$$
$$= \frac{\partial \delta u_i}{\partial x_r}\frac{\partial x_r}{\partial X_M} = \frac{\partial \delta u_i}{\partial x_r}F_{rM} \tag{6.22}$$

$$\frac{d}{d\epsilon}S_{MI}\bigg|_{\epsilon\to 0} = \frac{\partial S_{MI}}{\partial E_{KL}}\frac{dE_{KL}}{d\epsilon}\bigg|_{\epsilon\to 0} = \frac{\partial S_{MI}}{\partial E_{KL}}\frac{1}{2}\frac{d}{d\epsilon}\left(\left(F_{jK}F_{jL} - \delta_{KL}\right)\right)\bigg|_{\epsilon\to 0}$$
$$= \frac{\partial S_{MI}}{\partial E_{KL}}\frac{1}{2}\left[\frac{d}{d\epsilon}F_{jK}\bigg|_{\epsilon\to 0}F_{jL} + \frac{d}{d\epsilon}F_{jL}\bigg|_{\epsilon\to 0}F_{jK}\right]$$
$$= \frac{\partial S_{MI}}{\partial E_{KL}}\frac{1}{2}\left[\frac{\partial \delta u_j}{\partial X_K}F_{jL} + \frac{\partial \delta u_j}{\partial X_L}F_{jK}\right] = \frac{\partial S_{MI}}{\partial E_{KL}}sym\left(\frac{\partial \delta u_h}{\partial X_K}F_{hL}\right)$$
$$= \frac{\partial S_{MI}}{\partial E_{KL}}\left(\frac{\partial \delta u_h}{\partial X_K}F_{hL}\right) \tag{6.23}$$

where $\boldsymbol{E}$ is the Green-Lagrange strain. Equation (6.21) is rewritten as:

$$\frac{d}{d\epsilon}\int_{V_0}P_{iI}\frac{\partial \delta v_i}{\partial X_I}dV_0\bigg|_{\epsilon\to 0} = \int_{V_0}\frac{\partial \delta u_i}{\partial x_r}\tau_{rI}\frac{\partial \delta v_i}{\partial X_I}dV_0 + \int_{V_0}\frac{\partial \delta u_h}{\partial X_K}F_{iM}\,F_{hL}\frac{\partial S_{MI}}{\partial E_{KL}}\frac{\partial \delta v_i}{\partial X_I}dV_0$$
$$= \int_{V_0}\frac{\partial \delta u_i}{\partial x_r}\tau_{rI}\frac{\partial \delta v_i}{\partial X_I}dV_0 + \int_{V_0}\frac{\partial \delta u_h}{\partial x_p}\frac{\partial x_p}{\partial X_K}F_{iM}\,F_{hL}\frac{\partial S_{MI}}{\partial E_{KL}}\frac{\partial \delta v_i}{\partial x_q}\frac{\partial x_q}{\partial X_I}dV_0$$
$$= \int_{V_0}\frac{\partial \delta u_i}{\partial x_r}\tau_{rI}\frac{\partial \delta v_i}{\partial X_I}dV_0 + \int_{V_0}\frac{\partial \delta u_h}{\partial x_p}F_{pK}F_{iM}\,F_{hL}F_{qI}\frac{\partial S_{MI}}{\partial E_{KL}}\frac{\partial \delta v_i}{\partial x_q}dV_0$$
$$= \int_{V_0}\frac{\partial \delta u_i}{\partial x_r}\tau_{rI}\frac{\partial \delta v_i}{\partial X_I}dV_0 + \int_{V_0}\frac{\partial \delta u_h}{\partial x_p}F_{pK}F_{iM}\,F_{hL}F_{qI}\,2\frac{\partial S_{MI}}{\partial C_{KL}}\frac{\partial \delta v_i}{\partial x_q}dV_0 \tag{6.24}$$

The fourth order tensor, $2\frac{\partial S_{MI}}{\partial C_{KL}}$, has been introduced before as (Equation 3.108)

$$\mathbb{C}_{MIKL} = 2\frac{\partial S_{MI}}{\partial C_{KL}} \tag{6.25}$$

On the other hand, the transformation between the Kirchhoff stress tensor and the second Piola-Kirchhoff stress tensor is:

$$\boldsymbol{\tau} = \boldsymbol{F}\boldsymbol{S}\boldsymbol{F}^T \tag{6.26}$$

To calculate the rate of $\boldsymbol{\tau}$, we have:

$$\dot{\boldsymbol{\tau}} = \dot{\boldsymbol{F}}\boldsymbol{S}\boldsymbol{F}^T + \boldsymbol{F}\dot{\boldsymbol{S}}\boldsymbol{F}^T + \boldsymbol{F}\boldsymbol{S}\dot{\boldsymbol{F}}^T = \boldsymbol{L}.\boldsymbol{\tau} + \boldsymbol{F}\dot{\boldsymbol{S}}\boldsymbol{F}^T + \boldsymbol{\tau}.\boldsymbol{L}^T$$
$$= \boldsymbol{L}.\boldsymbol{\tau} + \boldsymbol{F}\left(\frac{\partial \boldsymbol{S}}{\partial \boldsymbol{E}}:\dot{\boldsymbol{E}}\right)\boldsymbol{F}^T + \boldsymbol{\tau}.\boldsymbol{L}^T \tag{6.27}$$

$\boldsymbol{L}$, the velocity gradient was introduced in Equation (4.2). Symmetric and unsymmetrical parts of $\boldsymbol{L}$ contribute to the definition of $\boldsymbol{D}$, rate of deformation gradient, and $\boldsymbol{W}$, spin tensor:

$$\boldsymbol{D} = \frac{1}{2}\left(\boldsymbol{L} + \boldsymbol{L}^T\right) \tag{6.28}$$

$$\boldsymbol{W} = \frac{1}{2}\left(\boldsymbol{L} - \boldsymbol{L}^T\right) \tag{6.29}$$

So, we have:

$$\dot{\boldsymbol{\tau}} - \left(\boldsymbol{W} + \boldsymbol{D}\right).\boldsymbol{\tau} - \boldsymbol{\tau}.\left(\boldsymbol{D} - \boldsymbol{W}\right) = \boldsymbol{F}\dot{\boldsymbol{S}}\boldsymbol{F}^T \tag{6.30}$$

$$\dot{\boldsymbol{\tau}} - \boldsymbol{W}.\boldsymbol{\tau} + \boldsymbol{\tau}.\boldsymbol{W} = \boldsymbol{F}\dot{\boldsymbol{S}}\boldsymbol{F}^T + \boldsymbol{D}.\boldsymbol{\tau} + \boldsymbol{\tau}.\boldsymbol{D} \tag{6.31}$$

The left-hand side of the relation (6.31) expresses the Jaumann rate of the Kirchhoff stress tensor:

$$\boldsymbol{\tau}^{\nabla(JK)} = \dot{\boldsymbol{\tau}} - \boldsymbol{W}.\boldsymbol{\tau} + \boldsymbol{\tau}.\boldsymbol{W} \tag{6.32}$$

Which leads to:

$$\boldsymbol{\tau}^{\nabla(JK)} = \boldsymbol{F}\dot{\boldsymbol{S}}\boldsymbol{F}^T + \boldsymbol{D}.\boldsymbol{\tau} + \boldsymbol{\tau}.\boldsymbol{D} \tag{6.33}$$

$$\boldsymbol{\tau}^{\nabla(JK)} = \boldsymbol{F}\left(\frac{\partial \boldsymbol{S}}{\partial \boldsymbol{E}} : \dot{\boldsymbol{E}}\right)\boldsymbol{F}^T + \boldsymbol{D}.\boldsymbol{\tau} + \boldsymbol{\tau}.\boldsymbol{D} = \boldsymbol{F}\left(\frac{\partial \boldsymbol{S}}{\partial \boldsymbol{E}}\boldsymbol{F}^T\boldsymbol{F}\boldsymbol{D}\right)\boldsymbol{F}^T + \boldsymbol{D}.\boldsymbol{\tau} + \boldsymbol{\tau}.\boldsymbol{D} \tag{6.34}$$

Finally, we have:

$$\begin{aligned}
\tau_{iq}^{\nabla(JK)} &= F_{pK}F_{iM}\,F_{hL}F_{qI}\frac{\partial S_{MI}}{\partial E_{KL}}D_{ph} + \frac{1}{2}\left(\tau_{ip}\delta_{qh} + \tau_{qh}\delta_{ip} + \tau_{ih}\delta_{qp} + \tau_{qp}\delta_{ih}\right)D_{ph} \\
&= 2F_{pK}F_{iM}\,F_{hL}F_{qI}\frac{\partial S_{MI}}{\partial C_{KL}}D_{ph} + \frac{1}{2}\left(\tau_{ip}\delta_{qh} + \tau_{qh}\delta_{ip} + \tau_{ih}\delta_{qp} + \tau_{qp}\delta_{ih}\right)D_{ph} \\
&= F_{pK}F_{iM}\,F_{hL}F_{qI}\mathbb{C}_{MIKL}D_{ph} + \frac{1}{2}\left(\tau_{ip}\delta_{qh} + \tau_{qh}\delta_{ip} + \tau_{ih}\delta_{qp} + \tau_{qp}\delta_{ih}\right)D_{ph}
\end{aligned} \tag{6.35}$$

Using Equation (3.112), we have:

$$\tau_{iq}^{\nabla(JK)} = J\,\mathbb{c}_{pihq}D_{ph} + \frac{1}{2}\left(\tau_{ip}\delta_{qh} + \tau_{qh}\delta_{ip} + \tau_{ih}\delta_{qp} + \tau_{qp}\delta_{ih}\right)D_{ph} \tag{6.36}$$

On the other hand, according to the Abaqus User Manual, the user has to define the expression for `DDSDDE` in UMAT, which is mentioned as follows:

$$\text{DDSDDE} = \frac{\partial \Delta \boldsymbol{\sigma}}{\partial \Delta \epsilon} = \frac{1}{J}\frac{\partial \Delta \boldsymbol{\tau}}{\partial \Delta \epsilon} \tag{6.37}$$

$\epsilon$ is known as STRAN in Abaqus updates according to:

$$\epsilon^{k+1} = \epsilon^{k} + \Delta\epsilon \tag{6.38}$$

where

$$\Delta\epsilon = \int_{t^{k+1}}^{t^{k+1}} \boldsymbol{D}\, dt \approx \boldsymbol{D} \cdot \Delta t \tag{6.39}$$

Abaqus User Manual also declares that Abaqus/Standard Abaqus/Explicit Solvers use Jaumann objective rate for Solid (Continuum) elements:

$$\Delta\boldsymbol{\tau} = \boldsymbol{\tau}^{\nabla(JK)} \cdot \Delta t \tag{6.40}$$

Therefore:

$$\text{DDSDDE} = \frac{1}{J}\frac{\partial \Delta\boldsymbol{\tau}}{\partial \Delta\epsilon} = \frac{1}{J}\frac{\boldsymbol{\tau}^{\nabla(JK)} \cdot \Delta t}{\boldsymbol{D} \cdot \Delta t} = \frac{1}{J}\frac{\boldsymbol{\tau}^{\nabla(JK)}}{\boldsymbol{D}} = \frac{1}{J}\frac{\boldsymbol{\tau}^{\nabla(JK)}}{\boldsymbol{D}} \tag{6.41}$$

Which leads to

$$\boldsymbol{\tau}_{iq}^{\nabla(JK)} = J\left(\text{DDSDDE}\right) : \boldsymbol{D} \tag{6.42}$$

Comparing Equations (6.36) and (6.42) reveals:

$$\begin{aligned} DDSDDE_{ipqh} &= \mathbb{c}_{pihq} + \frac{1}{2J}\left(\tau_{ip}\delta_{qh} + \tau_{qh}\delta_{ip} + \tau_{ih}\delta_{qp} + \tau_{qp}\delta_{ih}\right) \\ &= \frac{2}{J} F_{pK} F_{iM}\, F_{hL} F_{ql} \frac{\partial S_{MI}}{\partial C_{KL}} + \frac{1}{2}\left(\sigma_{ip}\delta_{qh} + \sigma_{qh}\delta_{ip} + \sigma_{ih}\delta_{qp} + \sigma_{qp}\delta_{ih}\right) \end{aligned} \tag{6.43}$$

In its most general form, consistent tangent modulus components, DDSDDE are arranged as:

$$DDSDDE = \begin{bmatrix} c_{1111} & c_{1122} & c_{1133} & c_{1112} & c_{1113} & c_{1123} \\ c_{2211} & c_{2222} & c_{2233} & c_{2212} & c_{2213} & c_{2223} \\ c_{3311} & c_{3322} & c_{3333} & c_{3312} & c_{3313} & c_{3323} \\ c_{1211} & c_{1222} & c_{1233} & c_{1212} & c_{1213} & c_{1223} \\ c_{1311} & c_{1322} & c_{1333} & c_{1312} & c_{1313} & c_{1323} \\ c_{2311} & c_{2311} & c_{2333} & c_{2312} & c_{2313} & c_{2323} \end{bmatrix} \tag{6.44}$$

Despite the rapid convergence [6], analytical derivation of consistent tangent modulus using Helmholtz free energy and the second Piola-Kirchhoff stress tensor is troublesome and prone to error. Some numerical procedures have been suggested. They are mainly perturbation techniques to develop forward difference approximations [3].

## 6.5 UMAT IMPLEMENTATION

This section is dedicated to an example constitutive law developed as a UMAT code. The model is proposed for a dielectric elastomer actuator reinforced by two fiber families. The free energy function is assumed as follows:

$$W = C_{10}\left(\bar{I}_1 - 3\right) + C_{01}\left(\bar{I}_2 - 3\right) + \frac{\kappa}{2}(J-1)^2 - \frac{1}{2}\varepsilon J\mathbf{C}^{-1} : (\mathbb{E} \otimes \mathbb{E})$$
$$+ \frac{k_1^{(1)}}{2k_2^{(1)}}\left\{\exp\left[k_2^{(1)}(I_4-1)^2\right]-1\right\} + \frac{k_1^{(2)}}{2k_2^{(2)}}\left\{\exp\left[k_2^{(2)}(I_6-1)^2\right]-1\right\} \quad (6.45)$$

Therefore, the model constants $C_{10}$, $C_{01}$, $\kappa$, $\varepsilon$, $k_1^{(1)}$, $k_2^{(1)}$, $k_1^{(2)}$, $k_2^{(2)}$, $A^{(1)}$, $A^{(2)}$ and electric field array $\mathbb{E}$, are introduced as "Props". Selected energy function (Equation 6.45) leads to the Cauchy stress tensor:

$$\sigma_{mn} = \frac{2}{J}F_{mi}\frac{\partial W}{\partial S_{ij}}F_{nj} = \frac{2}{J}\left(C_{10}\left(\bar{B}_{mn} - \frac{1}{3}\bar{I}_1\delta_{mn}\right) + C_{01}\left(\bar{I}_1\bar{B}_{mn} - \bar{B}_{mk}\bar{B}_{nk} - \frac{2}{3}\bar{I}_2\delta_{mn}\right)\right.$$
$$+\frac{\kappa}{2}(J-1)J\delta_{mn}\Bigg) + \varepsilon\left(\mathbb{e}_m\mathbb{e}_n - \frac{1}{2}\mathbb{e}_l\mathbb{e}_l\delta_{mn}\right)$$
$$+\frac{2}{J}\left(k_1^{(1)}(I_4-1)\left(\exp\left(k_2^{(1)}(I_4-1)^2\right)\right)a_m^{(1)}a_n^{(1)}\right.$$
$$\left.+k_1^{(2)}(I_6-1)\left(\exp\left(k_2^{(2)}(I_6-1)^2\right)\right)a_m^{(2)}a_n^{(2)}\right) \quad (6.46)$$

Then, the consistent tangent modulus is defined as:

$$\text{DDSDDE} = \frac{2}{J}F_{mi}F_{nj}F_{rk}F_{sl}\frac{\partial S_{ij}}{\partial C_{kl}} + \frac{1}{2}\left(\sigma_{mr}\delta_{ns} + \sigma_{ms}\delta_{nr} + \sigma_{nr}\delta_{ms} + \sigma_{ns}\delta_{mr}\right)$$
$$= \left(\frac{4}{3}\left(C_{10}\bar{I}_1 + 4C_{01}\bar{I}_2\right) + \kappa(2J-1)\right)\delta_{mn}\delta_{rs} + \left(2\bar{I}_1C_{10} + 4\bar{I}_2C_{01} + \kappa(1-J)\right)$$
$$\left(\delta_{mr}\delta_{ns} + \delta_{nr}\delta_{ms}\right) - 4\left(C_{10} + 2C_{01}\bar{I}_1\right)\left(\bar{B}_{mn}\delta_{rs} + \bar{B}_{rs}\delta_{mn}\right)$$
$$+4C_{01}\left(2\left(\bar{B}_{rk}\bar{B}_{sk}\delta_{mn} + \bar{B}_{mk}\bar{B}_{nk}\delta_{rs}\right) - 2\bar{B}_{mn}\bar{B}_{rs} + \left(\bar{B}_{mr}\bar{B}_{ns} + \bar{B}_{nr}\bar{B}_{ms}\right)\right)$$
$$-\frac{1}{2}\varepsilon\left(\left(\delta_{mn}\delta_{rs} - \left(\delta_{mr}\delta_{ns} + \delta_{ms}\delta_{nr}\right)\right)\mathbb{e}_l\mathbb{e}_l\right)$$
$$+\varepsilon\left(\mathbb{e}_n\mathbb{e}_m\delta_{rs} + \mathbb{e}_r\mathbb{e}_s\delta_{mn} - \mathbb{e}_n\mathbb{e}_s\delta_{mr} - \mathbb{e}_m\mathbb{e}_r\delta_{ns} - \mathbb{e}_m\mathbb{e}_s\delta_{nk} - \mathbb{e}_n\mathbb{e}_r\delta_{ms}\right)$$
$$+\frac{4}{J}\left(k_1^{(1)}\left(1 + 2k_2^{(1)}(I_4-1)^2\right)\exp\left(k_2^{(1)}(I_4-1)^2\right)a_m^{(1)}a_n^{(1)}a_r^{(1)}a_s^{(1)}\right.$$
$$\left.+k_1^{(2)}\left(1 + 2k_2^{(2)}(I_6-1)^2\right)\exp\left(k_2^{(2)}(I_6-1)^2\right)a_m^{(2)}a_n^{(2)}a_r^{(2)}a_s^{(2)}\right)$$
$$+\frac{1}{2}\left(\sigma_{mr}\delta_{ns} + \sigma_{ms}\delta_{nr} + \sigma_{nr}\delta_{ms} + \sigma_{ns}\delta_{mr}\right) \quad (6.47)$$

Users may prefer to define multipliers like $\kappa(2J-1)$ as frequently used constants.

Therefore, the user first needs to determine terms of $J$, $\bar{B}_{mn}$, $\bar{I}_1$, $\bar{I}_2$, $e_m$, $I_4$, $I_6$, $a_m^{(1)}$, $a_m^{(2)}$. In this regard, $F$ is "DFGRD0" and is used to define $J$ (Equation 3.8), $\bar{\boldsymbol{F}}$ (Equation 3.63), $\bar{\boldsymbol{C}}$ (Equation 3.64), $\bar{I}_1$ and $\bar{I}_2$ (Equation 3.68 and 3.69), $I_4$, $I_6$, $a^{(1)}$ and $a^{(2)}$ (Equation 3.85), $e$ (Equation 3.100).

Once these variables are defined, the user has to develop the components of "`STRESS`" and "`DDDSDDE`" according to (6.46) and (6.47). It is worth noting that tensors are defined in Voigt notation; i.e. "`STRESS`" and "`DDSDDE`" are mentioned as first and second order tensors.

This way, the UMAT development is complete. To implement this UMAT in a simulation, the user has to select "User Material" under "General" tab when defining a material in "Property" module. Then "Data" option is used to provide the model constants as inputs. To include solution-dependent state variables, the user selects "Depvar" under "General" tab and then determines the number of "`STATEV`". This procedure allocates proper storage to record these variables. The state variables can be requested as output "SVD" via "Field Output Manager" under "Step" module.

At this point the UMAT and the simulation file are developed, however, they are not related. To tell the solver to use the UMAT as the material definition, the user gives its path when creating the "Job".

When the "Job" is submitted, The Abaqus/Standard solver reads the subroutine and updates the "`STRESS`" and "`DDSDDE`" for each iteration to the last one.

## 6.6 CONCLUSION

Chapters 3 and 4 presented a mathematical model to describe different characteristics of a dielectric elastomer actuator. The properties of the model and how it was developed makes it appropriate to be used as a numerical algorithm via finite element scheme and softwares. Therefore, Chapter 6 devised the procedure of numerical implementation of the model. This chapter was dedicated to create a UMAT subroutine from the developed constitutive law and use it for Abaqus. Therefore, it first introduced the UMAT subroutine, its features, necessary parts, and parameters. Then, Cauchy stress and tangent modulus tensor components were expressed. The results of numerical analysis will be available in Chapter 8.

## REFERENCES

[1] Hibbitt, Karlsson, & Sorensen. (2002). *ABAQUS/CAE user's manual.* Hibbitt, Karlsson & Sorensen, Incorporated.

[2] J. Bonet, & R. D. Wood, *Nonlinear continuum mechanics for finite element analysis.* Cambridge University Press, 1997.

[3] C. Miehe, Numerical computation of algorithmic (consistent) tangent moduli in large-strain computational inelasticity. *Computer Methods in Applied Mechanics and Engineering* 134(3–4) (1996) 223–240.

[4] N. Nguyen, & A. M. Waas, Nonlinear, finite deformation, finite element analysis. *Zeitschrift für angewandte Mathematik und Physik* 67 (2016) 1–24.

[5] Hibbitt, Karlsson, & Sorensen. (1997). *ABAQUS: Theory manual* (Vol. 2). Hibbitt, Karlsson & Sorensen.

[6] W. Sun, E. L. Chaikof, & M. E. Levenston, Numerical approximation of tangent moduli for finite element implementations of nonlinear hyperelastic material models. *Journal of Biomechanical Engineering*, 130 (2008) 061003.

# Part III

## Applications and Implementation

# 7 Practical Applications of Multi-Layered Fiber-Reinforced Dielectric Elastomer Composites

## 7.1 INTRODUCTION

Previously, in Chapter 2 it was mentioned that increasing the number of dielectric layers while the total thickness is constant, decreases the actuation voltage [1–3]. Therefore, multilayer dielectric elastomers produce higher actuation strains and output forces. Lower actuation voltage also leads to more stable performance.

On the other hand, fiber-reinforced dielectric elastomer actuators benefit from more considerable actuation deformation and electromechanical stability. Fibers also provide anisotropic behavior and deformation which suggest more application for DEAs.

Therefore, a fiber-reinforced multi-layered dielectric elastomer actuator seems promising to promote potential application for a DEA.

Motivated by this idea, Chapter 7 is dedicated to multi-layered fiber-reinforced dielectric elastomer composites. Firstly, some practical applications reported in previous studies are reviewed. Then a framework to study fiber-reinforced multi-layered dielectric elastomer is proposed. It is mainly based on the introduced model in previous chapters with some modifications to include the effect of additional layers.

## 7.2 MULTI-LAYERED ACTUATORS FOR SOFT ROBOTS

Despite promising findings of fiber-reinforced multi-layered dielectric elastomer actuators, investigations on their practical applications are limited, mainly because of technical difficulties of fabrication process. However, studies on the soft robots employing multi-layered dielectric elastomer actuators have reported acceptable performance. This section presents some of these studies.

### 7.2.1 Unimorph and Bimorph Actuator

Multi-layered structures provide various characteristics because of the opportunity to employ materials with different features. A familiar structure which benefits

DOI: 10.1201/9781003571469-10

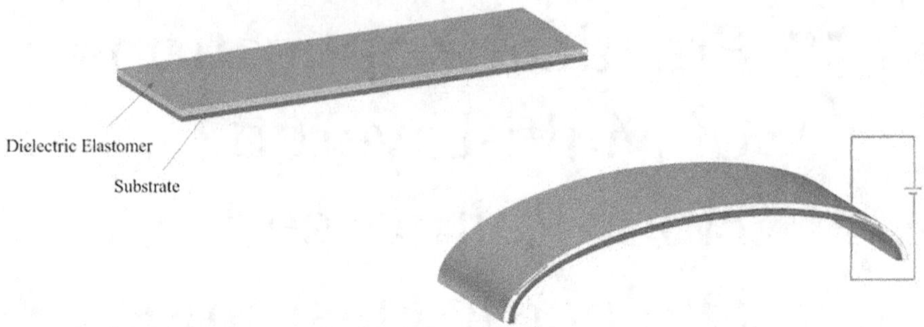


**FIGURE 7.1** A unimorph actuator structure and its actuated state.

from such an opportunity are unimorph and bimorph dielectric elastomer actuators. Stacking single or multi-layered dielectric elastomer actuators on inactive substrate creates unimorph and bimorph actuator which are able to perform large out-of- plane deformations. Figures 7.1 and 7.2 depict schematic presentations of a unimorph and bimorph structures.

The substrate is flexible on bending and stiff in tension [4]. Therefore, as an electric field is applied to the dielectric elastomer, the substrate constraints in-plane deformation and induce bending. Properties of the elastomer and substrate, the structure geometry, in addition to the number of layers determine bending deformation limits. Induced deformation stores elastic energy. This structure has been reported to devise steerable endoscopes [4].

The biaxial response of an isotropic unimorph actuator, in cases where the actuator is supposed to mimic biological muscles, is not desirable. Therefore, strategies to achieve unidirectional deformation and optimize efficiency are of interest. Embedding aligned fibers perpendicular to the desired direction for the

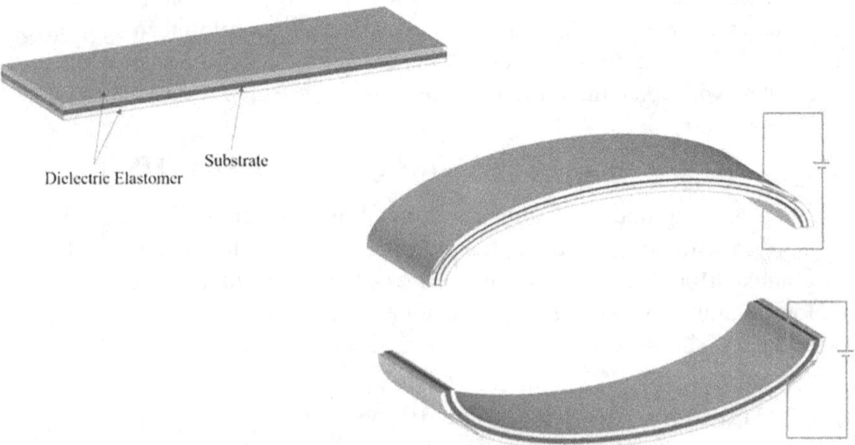


**FIGURE 7.2** A bimorph actuator structure and its actuated states.

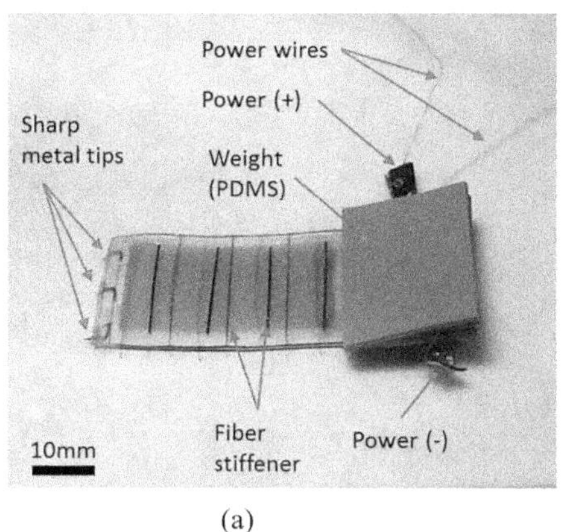


(a)

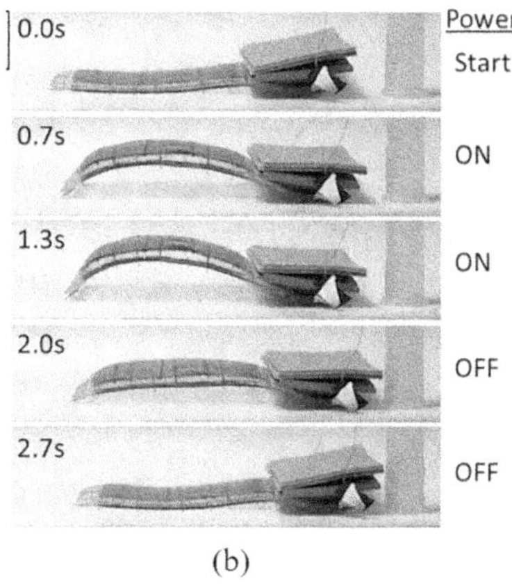


(b)

**FIGURE 7.3** (a) Inchworm robot (b) Walking movement [5].

deformation is one of the strategies to improve actuation performance. It was used by shian et. al. [5] to develop an inchworm robot, according to Figure 7.3(a). They reinforced three layers of VHB films with a family of nylon fibers to improve the robot's performance. Successive contraction and relaxation cycles of the VHB layers caused by actuation signals change the net body length: application of the electric excitation makes the body bend. So, the tail of the robot moves forward. The excitation vanishes, the dielectric film relaxes, and the head proceeds. By the end of this step, the robot attains its initial length while it has moved forward. It worth mention that anisotropic friction enables movement of the head and tail to happen one by one so the robot can dislocate. Figure 7.3(b) shows the inchworm robot movement in one cycle.

The study also revealed the effect of fiber reinforcement on the (undesired) lateral contraction.

A structure similar to the bimorph depicted in Figure 7.2 which has a hyper-elastic layer instead of the elastic substrate was introduced and theoretically studied in Ref. [6]. The DE layers are reinforced with fibers in order to reduce the derive voltage and bending deformation capacity. It performs as a bidirectional bending actuator.

### 7.2.2 Moving Robots

Issues including viscoelasticity, fluid rate and complex infrastructure usually limit the application of pneumatic and hydraulic actuators to design walking robots. Therefore, a potential alternative for such systems are dielectric elastomer actuators because they are soft and respond to the electric excitation, fast. Moreover, reduced thickness of the active layers in multi-layered dielectric elastomer actuators is observed to decrease the activation voltage sufficiently. So, multi-layered DEAs were used to devise a high-speed soft robot [7]. The multi-layered (five active layers) DEAs are four unimorph actuators as the robot's legs, as depicted in Figure 7.4.

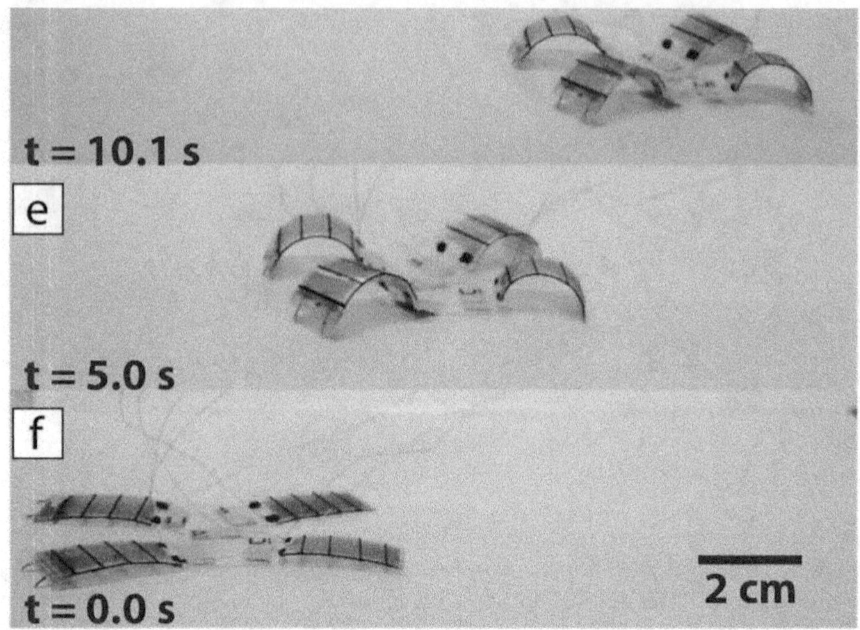


**FIGURE 7.4** The four-legged walking robot actuated at 1.5 Hz [7].

As the reinforced unimorph actuators bend and attain their initial flat form, the robot moves forward. The unimorph actuators used as the legs are also reinforced by some carbon fibers to control the actuation direction.

The other moving robot employing multi-layered DEAs is reported by Nguyen et al. [8]. It is a quadruped robot with four 2DOFs legs. Each leg of the robot mimics the performance of hip and knee joints in a two-phase operation.

The leg performs walking as the two multi-layered DEAs (a pile of electrically parallel dielectric elastomer films) are actuated (swing phase) and relaxed (stance phase) by turn. In swing phase the foot (lowest point of the leg) lifts and as it touches the ground at the end of stance phase it has moved forward for 34 mm. Linear motions of the actuators are converted to rotational ones using slider crank mechanisms. The motion process is depicted in Figure 7.5(a). Four legs assembled into a body led to a quadruped robot Figure 7.5(b). It weighed 450 g.

Multi-layered dielectric elastomer actuators are also proposed to design jumping robots [9]. According to the reported findings for a force-voltage test device designed to measure force-voltage relationship, output force produced by a single-layer DEA slightly changes when the voltage increases from 4 to 7 kV. As the number of layers increase, voltage variations affect output force stronger. Increasing the number of layers from one to 20, not only leads to 15 times more significant output force for a constant voltage at 7 kV, it causes the steeper increase in output force (Figure 7.6(a)).

Encouraged by this result, the authors proposed and fabricated a jumping robot, shown in Figure 7.6(b). The robot is designed based on a structure consisting a

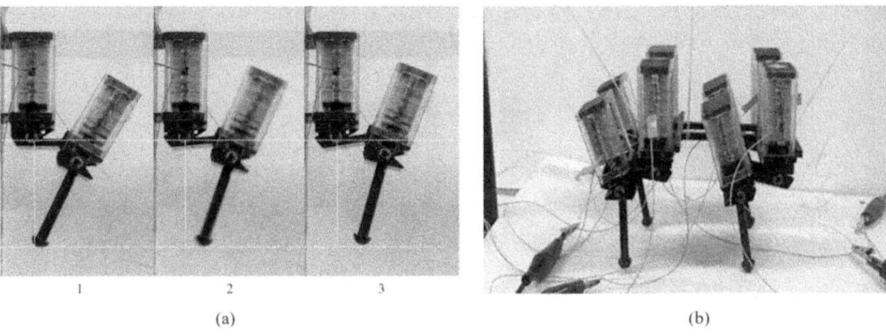


**FIGURE 7.5** (a) A walking leg (b) Assembled quadruped robot [8].

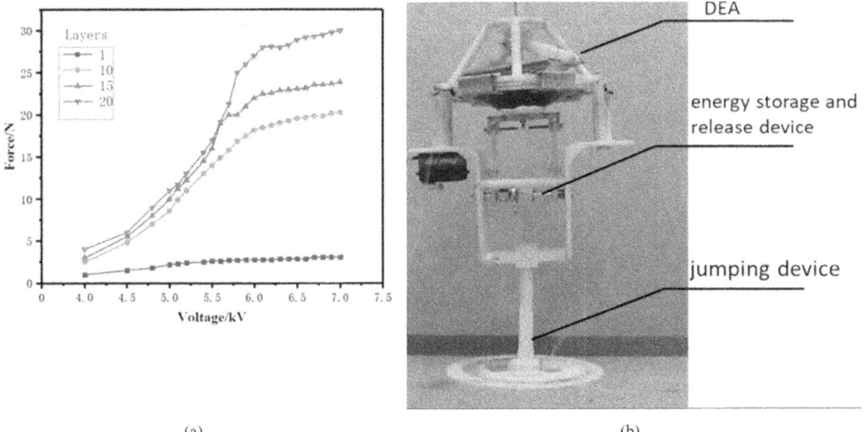


**FIGURE 7.6** (a) Voltage-output force for different number of DE layers (b) Jumping robot overall prototype [9].

multi-layered dielectric elastomer actuator (to produce the driving force), a jumping device and an energy storage and release device. The designers used multilayer actuator because higher output force leads to increased jumping height. Electric excitation required to actuate the actuator is provided using a power supply device. It supplies a square wave voltage of different frequencies in different phase of the robot performance. The robot is 200 mm high and is able to jump up to 45 mm.

### 7.2.3 Biocompatible Devices

The reduction of activation voltage due to layering up enhances the potential of DEAs for biomedical applications. Therefore, multi-layered DEAs are proposed to design a pumping micromixer [10]. According to Figure 7.7, the device is made of pumping and mixing chambers and the chambers consists of some stack actuators. It can be seen that the walls of chambers are able to move as actuators are actuated. Therefore, the micromixer can mix and pump the fluid. Periodic motions

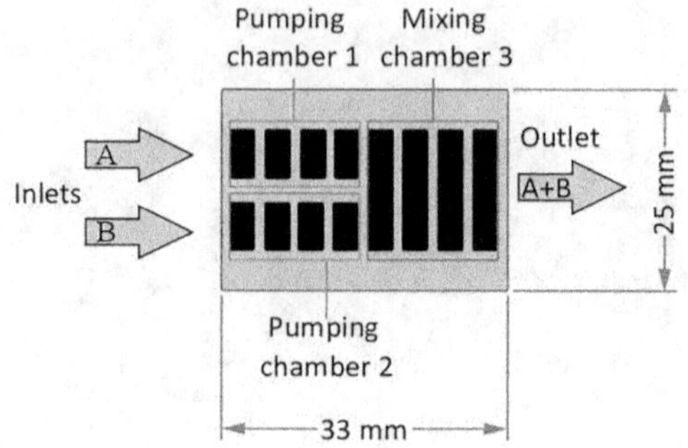


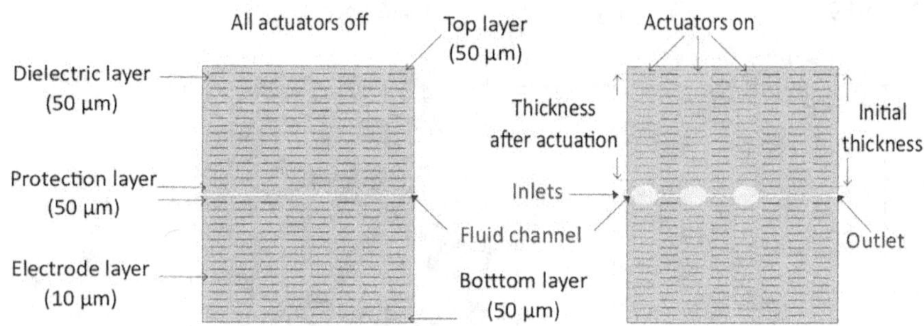


**FIGURE 7.7** Micromixer configuration [10].

of the walls are transferred to the fluids and mix them. The mixing ratio of the two solutions and flow rate can be controlled by the actuation frequency. Moreover, the micromixer is small enough and biocompatible.

### 7.2.4 Vibrotactile Display

Dielectric elastomer transducers may perform as actuator or sensor. This dual functionality makes them good choices to produce vibrotactile displays as an interface device. On the other hand, since the user is supposed to touch the interface directly, it has to be safe enough, which is provided using reduced activation voltage in multi-layered dielectric elastomer transducer. Such application is reported by Matyesk et al. [11]. They used a stack of 50 DE films to create a vibrotactile display to control the functionality of a multimedia device. Application of patterned electrode enable the deriving voltage to actuate particular regions and present operating conditions in form of haptic information. Similar mechanism was previously used to design and fabricate refreshable brail display [12].

### 7.2.5 Gas Valve

Since the application of electrostatic forces is the basics for many small-size actuation devices, dielectric elastomer stack actuators are frequently reported to fabricate valves, especially because of material softness (to address the need for sealing);

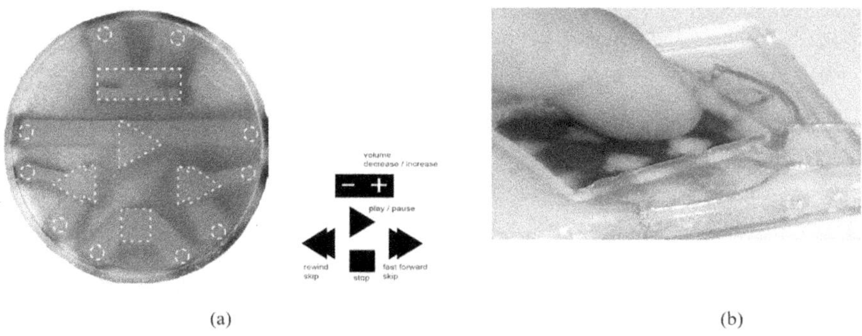


**FIGURE 7.8** (a) Vibrotactile display (b) Refreshable brail display [11].

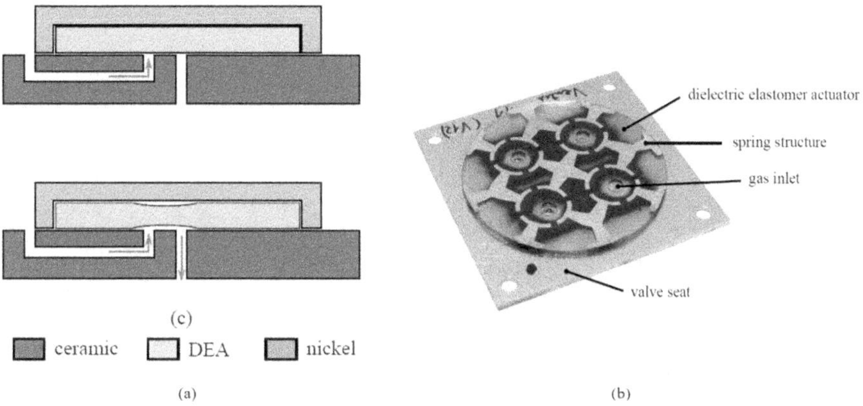


**FIGURE 7.9** (a) Valve structure configuration (b) Valve array [9].

i.e., reference [9] proposed a gas valve made of a dielectric actuator on a valve seat with a 2 by 2 valve array. The design includes a stack actuator pressed against the valve seat by a nickel spring, as in Figures 7.8 and 7.9. Geometrical arrangement of the stack actuator was optimized to achieve a predefined deflection.

In the absence of electric excitation, the metallic spring keeps the actuator in its place on the valve seat and closes the outlet. As the actuator is derived by the electric field, its total thickness decreases and lets the fluid flow.

## 7.3 GOVERNING EQUATIONS FOR MULTI-LAYERED FIBER-REINFORCED DIELECTRIC ELASTOMERS

To model a multilayer dielectric elastomer actuator, we assume that the sample consists of $\mathbb{N}$ elastomer layers, and each layer has an initial thickness $d_{(m)}$, where $m = 1,\ldots,\mathbb{N}$. For each layer, we define a local coordinate system at its bottom level. Figure 7.10 shows a schematic representation of the multilayer structure and the $m$th layer.

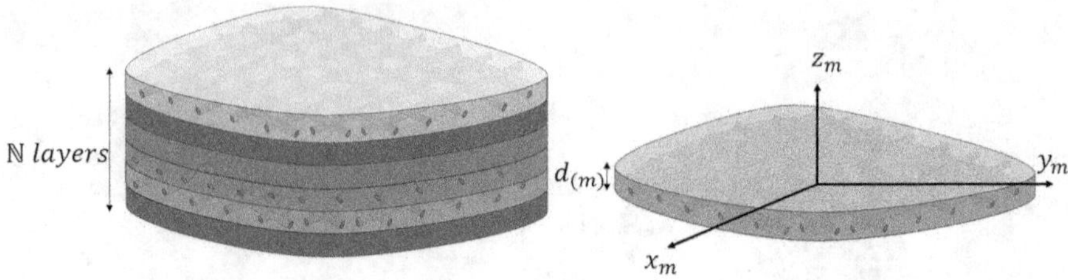


**FIGURE 7.10** Reference configuration of a fiber-reinforced multi-layered dielectric elastomer actuator and the *m*th layer configuration.

Application of mechanical loading and electrical excitation cause the layers to change shape. As a result, a deformation gradient is defined for each single layer, where $\boldsymbol{x}$ and $\boldsymbol{X}$ represent the current vectors and the reference position, respectively.

$$\boldsymbol{F}_{(m)} = \frac{d\boldsymbol{x}_{(m)}}{d\boldsymbol{X}_{(m)}} \tag{7.1}$$

Each layer may have different properties and characteristics depending on its material. Therefore, the model used to describe this actuator must be able to consider the unique characteristics of each layer; for this purpose, the free energy and stress are calculated for each layer. These parameters for the *m*th layer are:

$$\begin{aligned} W_{(m)} &= W_{(m)}^{\text{hyperelastic}} + W_{(m)}^{\text{anisotripic}} + W_{(\mathrm{m})}^{\text{dielectric}} = C_{10(m)}\left(I_{1(m)}-3\right) + C_{01(m)}\left(I_{2(m)}-3\right) \\ &+ \frac{k_{1(m)}^{(1)}}{2k_{2(m)}^{(1)}}\left\{\exp\left[k_{2(m)}^{(1)}\left(I_{4(m)}-1\right)^2\right]-1\right\} + \frac{k_{1(m)}^{(2)}}{2k_{2(m)}^{(2)}}\left\{\exp\left[k_{2(m)}^{(2)}\left(I_{6(m)}-1\right)^2\right]-1\right\} \\ &- \frac{1}{2}\varepsilon_{(m)}J_{(m)}\boldsymbol{C}_{(m)}^{-1} : \left[\mathbb{E}_{(m)}\mathbb{E}_{(m)}\right], m = 1,\ldots,\mathbb{N} \end{aligned} \tag{7.2}$$

$$\begin{aligned} \sigma_{(m)} &= \frac{2}{J_{(m)}}\left(C_{10(m)}\left(-\frac{1}{3}\bar{I}_{1(m)}\boldsymbol{I} + \bar{\boldsymbol{B}}_{(m)}\right) + C_{01(m)}\left(-\frac{2}{3}\bar{I}_{2(m)}\boldsymbol{I} + \left(\bar{I}_1\bar{\boldsymbol{B}}_{(m)} - \bar{\boldsymbol{B}}_{(m)}\bar{\boldsymbol{B}}_{(m)}\right)\right)\right) \\ &+ \kappa_{(m)}\left(J_{(m)}-1\right)\boldsymbol{I} + \varepsilon\left(\mathbb{e}_{(m)}\otimes\mathbb{e}_{(m)} - \frac{1}{2}\left(\mathbb{e}_{(m)}\cdot\mathbb{e}_{(m)}\right)\boldsymbol{I}\right) \\ &+ \frac{2}{J_{(m)}}k_{1(m)}^{(1)}\left(I_{4(m)}-1\right)\exp\left[k_{2(m)}^{(1)}\left(I_{4(m)}-1\right)^2\right]\boldsymbol{a}_{(m)}^{(1)}\otimes\boldsymbol{a}_{(m)}^{(1)} + \frac{2}{J_{(m)}}k_{1(m)}^{(2)}\left(I_{6(m)}-1\right) \\ &\exp\left[k_{2(m)}^{(2)}\left(I_{6(m)}-1\right)^2\right]\boldsymbol{a}_{(m)}^{(2)}\otimes\boldsymbol{a}_{(m)}^{(2)}, m = 1,\ldots,\mathbb{N} \end{aligned} \tag{7.3}$$

Therefore, the strain energy of the whole multi-layered structure is defined as follows:

$$W = \sum_{m=1}^{\mathbb{N}} v_{(m)} W_{(m)} \tag{7.4}$$

$v_{(m)}$ and $W_{(m)}$ are the volume fraction and strain energy of the *m*th layer, respectively. In addition, the following equation defines the Cauchy stress of the multi-layered structure:

$$\sigma = \sum_{m=1}^{\mathbb{N}} v_{(m)} \sigma_{(m)} \tag{7.5}$$

This equation should be rewritten according to the specific conditions of the structure. For example, if a layer is not fiber-reinforced or electrically active, the terms for that property will be omitted from the stress expression. Also, the constants of the models in each layer are specified according to the type of layer.

## 7.4 CONCLUSION

Parts I and II tried to declare the leading features of DEAs, their cons and pros, a mathematical model to study their behavior and its numerical implementation. Chapter 7 was mainly dedicated to fiber-reinforced multi-layered dielectric elastomer actuators. It reviewed the limited number of reported experimental analyses. Considering the few cases, encourages a mathematical model and numerical framework to investigate the behavior of such an actuator in advance. This model is a cost-effective tool to not only investigate a potential anisotropic multi-layered actuator, but also to optimize it for the maximum efficiency and other possible targets. Therefore, the single and multi-layered model would be employed and the numerical results for various dielectric elastomer actuators would be presented in the final chapter.

## REFERENCES

[1] Lotz, P., Matysek, M., Schlaak, H. F., Fabrication and application of miniaturized dielectric elastomer stack actuators, *IEEE/ASME Transactions on Mechatronics*, 16(1) (2010) 58–66.

[2] Duduta, M., Wood, R. J., Clarke, D. R., Multilayer dielectric elastomers for fast, programmable actuation without prestretch, *Advanced Materials*, 28(36) (2016) 8058–8063.

[3] Poulin, A., Rosset, S., Shea, H. R., Printing low-voltage dielectric elastomer actuators, *Applied Physics Letters*, 107(24) (2015) 244104.

[4] Araromi, O. A., Burgess, S. C., A finite element approach for modelling multilayer unimorph dielectric elastomer actuators with inhomogeneous layer geometry, *Smart Materials and Structures*, 21(3) (2012) 032001.
[5] Shian, S., Bertoldi, K., Clarke, D. R., Use of aligned fibers to enhance the performance of dielectric elastomer inchworm robots, in: Electroactive Polymer Actuators and Devices (EAPAD) 2015, SPIE, 2015, April, Vol. 9430, pp. 417–425.
[6] He, L., Lou, J., Du, J., Wang, J., Finite bending of a dielectric elastomer actuator and pre-stretch effects, *International Journal of Mechanical Sciences* 122 (2017) 120–128.
[7] Duduta, M., Clarke, D. R., Wood, R. J. A high speed soft robot based on dielectric elastomer actuators, in: 2017 IEEE International Conference on Robotics and Automation (ICRA) IEEE 2017, May, pp. 4346–4351.
[8] Nguyen, C. T., Phung, H., Nguyen, T. D., Lee, C., Kim, U., Lee, D., … Choi, H. R., A small biomimetic quadruped robot driven by multistacked dielectric elastomer actuators. *Smart Materials and Structures* 23(6) (2014) 065005.
[9] Flittner, K., Schlosser, M., Schlaak, H. F., Dielectric elastomer stack actuators for integrated gas valves, in: *Electroactive Polymer Actuators and Devices (EAPAD) 2011*, SPIE, 2011, March, Vol. 7976, pp. 443–449.
[10] Solano-Arana, S., Klug, F., Mößinger, H., Förster-Zügel, F., Schlaak, H. F., A novel application of dielectric stack actuators: A pumping micromixer. *Smart Materials and Structures*, 27(7) (2018) 074008.
[11] Matysek, M., Lotz, P., Flittner, K., Schlaak, H. F., Vibrotactile display for mobile applications based on dielectric elastomer stack actuators, in: *Electroactive Polymer Actuators and Devices (EAPAD) 2010*, SPIE, 2010, April, Vol. 7642, pp. 83–91.
[12] Matysek, M., Lotz, P., Schlaak, H. F., Tactile display with dielectric multilayer elastomer actuators, in: *Electroactive Polymer Actuators and Devices (EAPAD) 2009*, SPIE, 2009, April, Vol. 7287, pp. 425–433.

# 8 Numerical Implementation of Practical Case Studies

## 8.1 INTRODUCTION

Part II of this book introduced a hyperplastic framework to develop mathematical models and study the behavior of a DE actuator. Then, Chapter 7 introduced some of the most famous applications of multi-layered dielectric elastomer actuators. Motivated by the reported findings and possible applications for these actuators, we implemented the developed model to conduct some numerical analysis via Abaqus.

Numerical results enable us to explore new possibilities and evaluate the performance of the actuators in a cost-effective and realistic enough way. It provides the opportunity to study the effect of various factors on the behavior of an actuator and to optimize it to address particular aims regarding existing constraints.

Understanding of the behavior of fiber-reinforced multi-layered DE actuators will lead to more efficient and promising designs. It reveals limitations and appropriate solutions. The prognosis on the actuator features increases the chance for new applications and potentials.

This chapter presents some of the numerical simulations and results. Firstly, we review the required expressions to develop the model. This section concludes equations reported in previous chapters, followed by calibration for the model constants and validation. Then we introduce analyzed actuators. The presented cases include multi-layered fiber-reinforced dielectric elastomer actuators in different geometries and conditions. This section is followed by the final part, which is the results.

## 8.2 MODEL DEVELOPMENT

Chapter 6 detailed the requirements to implement a UMAT code based on the developed constitutive law of the material. Therefore, the expressions for Cauchy stress and consistent tangent modulus components.

Here, we assumed the energy function as follows for a fiber-reinforced dielectric elastomer actuator:

$$
\begin{aligned}
W = {} & C_{10}\left(\bar{I}_1 - 3\right) + C_{01}\left(\bar{I}_2 - 3\right) + \frac{\kappa}{2}(J-1)^2 - \frac{1}{2}\varepsilon J C^{-1} : (\mathbb{E} \otimes \mathbb{E}) \\
& + \frac{k_1^{(1)}}{2k_2^{(1)}}\left\{\exp\left[k_2^{(1)}(I_4 - 1)^2\right] - 1\right\} + \frac{k_1^{(2)}}{2k_2^{(2)}}\left\{\exp\left[k_2^{(2)}(I_6 - 1)^2\right] - 1\right\}
\end{aligned} \tag{8.1}
$$

DOI: 10.1201/9781003571469-11

This function includes the hyperelasticity, incompressibility, and dielectric effect of the dielectric elastomer matrix. It also covers the stiffening effect of two fiber families. The model constants $C_{10}$, $C_{01}$, $\kappa$, $\varepsilon$, $k_1^{(1)}$, $k_2^{(1)}$, $k_1^{(2)}$, $k_2^{(2)}$, $A^{(1)}$, $A^{(2)}$ need to be defined for the material. We used the experimental results to determine the model constants or calibrate the model. Calibration will be explained later. The electric field array $\mathbb{E}$, will be determined considering the simulation and its condition.

The strain energy function is derived and pushed forward to develop the Cauchy stress components:

$$\begin{aligned}\sigma_{mn}^{e} &= \frac{2}{J}F_{mi}\frac{\partial W}{\partial S_{ij}}F_{nj} = \frac{2}{J}\left(C_{10}\left(\bar{B}_{mn} - \frac{1}{3}\bar{I}_1\delta_{mn}\right)\right. \\ &+ C_{01}\left(\bar{I}_1\bar{B}_{mn} - \bar{B}_{mk}\bar{B}_{nk} - \frac{2}{3}\bar{I}_2\delta_{mn}\right) + \frac{\kappa}{2}(J-1)J\delta_{mn}\Bigg) \\ &+ \varepsilon\left(\mathbb{e}_m\mathbb{e}_n - \frac{1}{2}\mathbb{e}_l\mathbb{e}_l\delta_{mn}\right) + \frac{2}{J}\left(k_1^{(1)}(I_4-1)\left(\exp\left(k_2^{(1)}(I_4-1)^2\right)\right)a_m^{(1)}a_n^{(1)}\right. \\ &\left. + k_1^{(2)}(I_6-1)\left(\exp\left(k_2^{(2)}(I_6-1)^2\right)\right)a_m^{(2)}a_n^{(2)}\right) \end{aligned} \tag{8.2}$$

In this expression, the electric field and fiber unit vectors are pushed to Eulerian form. The equation to transform these vectors are presented in previous chapters and are included in the UMAT code. The UMAT code also requires a consistent tangent modulus as:

$$\begin{aligned} DDSDDE &= \frac{2}{J}F_{mi}F_{nj}F_{rk}F_{sl}\frac{\partial S_{ij}}{\partial C_{kl}} + \frac{1}{2}\left(\sigma_{mr}\delta_{ns} + \sigma_{ms}\delta_{nr} + \sigma_{nr}\delta_{ms} + \sigma_{ns}\delta_{mr}\right) \\ &= \left(\frac{4}{3}\left(C_{10}\bar{I}_1 + 4C_{01}\bar{I}_2\right) + \kappa(2J-1)\right)\delta_{mn}\delta_{rs} + \left(2\bar{I}_1C_{10} + 4\bar{I}_2C_{01} + \kappa(1-J)\right) \\ &\quad \left(\delta_{mr}\delta_{ns} + \delta_{nr}\delta_{ms}\right) - 4\left(C_{10} + 2C_{01}\bar{I}_1\right)\left(\bar{B}_{mn}\delta_{rs} + \bar{B}_{rs}\delta_{mn}\right) \\ &\quad + 4C_{01}\left(2\left(\bar{B}_{rk}\bar{B}_{sk}\delta_{mn} + \bar{B}_{mk}\bar{B}_{nk}\delta_{rs}\right) - 2\bar{B}_{mn}\bar{B}_{rs} + \left(\bar{B}_{mr}\bar{B}_{ns} + \bar{B}_{nr}\bar{B}_{ms}\right)\right) \\ &\quad - \frac{1}{2}\varepsilon\left(\left(\delta_{mn}\delta_{rs} - \left(\delta_{mr}\delta_{ns} + \delta_{ms}\delta_{nr}\right)\right)\mathbb{e}_l\mathbb{e}_l\right) \\ &\quad + \varepsilon\left(\mathbb{e}_n\mathbb{e}_m\delta_{rs} + \mathbb{e}_r\mathbb{e}_s\delta_{mn} - \mathbb{e}_n\mathbb{e}_s\delta_{mr} - \mathbb{e}_m\mathbb{e}_r\delta_{ns} - \mathbb{e}_m\mathbb{e}_s\delta_{nk} - \mathbb{e}_n\mathbb{e}_r\delta_{ms}\right) \\ &\quad + \frac{4}{J}\left(k_1^{(1)}\left(1 + 2k_2^{(1)}(I_4-1)^2\right)\exp\left(k_2^{(1)}(I_4-1)^2\right)a_m^{(1)}a_n^{(1)}a_r^{(1)}a_s^{(1)}\right. \\ &\quad \left. + k_1^{(2)}\left(1 + 2k_2^{(2)}(I_6-1)^2\right)\exp\left(k_2^{(2)}(I_6-1)^2\right)a_m^{(2)}a_n^{(2)}a_r^{(2)}a_s^{(2)}\right) \\ &\quad + \frac{1}{2}\left(\sigma_{mr}\delta_{ns} + \sigma_{ms}\delta_{nr} + \sigma_{nr}\delta_{ms} + \sigma_{ns}\delta_{mr}\right) \end{aligned} \tag{8.3}$$

The procedure to derive these expressions are presented in details in Part II.

Including the rate-dependency feature modifies the free energy function, Cauchy stress, and consistent tangent modulus as Equations (7.4) to (7.6):

$$
\begin{aligned}
W &= C_{10}\left(\bar{I}_1-3\right)+C_{01}\left(\bar{I}_2-3\right)+\frac{\kappa}{2}(J-1)^2-\frac{1}{2}\varepsilon J\boldsymbol{C}^{-1}:(\mathbb{E}\otimes\mathbb{E}) \\
&+\frac{k_1^{(1)}}{2k_2^{(1)}}\left\{\exp\left[k_2^{(1)}\left(I_4-1\right)^2\right]-1\right\}+\frac{k_1^{(2)}}{2k_2^{(2)}}\left\{\exp\left[k_2^{(2)}\left(I_6-1\right)^2\right]-1\right\} \\
&+\frac{1}{4}\left(\bar{I}_1-3\right)\left(\eta_1\bar{J}_2+\eta_2\bar{J}_4^2\right)
\end{aligned}
\tag{8.4}
$$

$$
\begin{aligned}
\sigma_{mn}^{e} &= \frac{2}{J}\left(C_{10}\left(\bar{B}_{mn}-\frac{1}{3}\bar{I}_1\delta_{mn}\right)+C_{01}\left(\bar{I}_1\bar{B}_{mn}-\bar{B}_{mk}\bar{B}_{nk}-\frac{2}{3}\bar{I}_2\delta_{mn}\right)+\frac{\kappa}{2}(J-1)J\delta_{mn}\right) \\
&+\varepsilon\left(\mathrm{e}_m\mathrm{e}_n-\frac{1}{2}\mathrm{e}_l\mathrm{e}_l\delta_{mn}\right) \\
&+\frac{2}{J}\left(k_1^{(1)}\left(I_4-1\right)\left(\exp\left(k_2^{(1)}\left(I_4-1\right)^2\right)\right)a_m^{(1)}a_n^{(1)}\right. \\
&\left.+k_1^{(2)}\left(I_6-1\right)\left(\exp\left(k_2^{(2)}\left(I_6-1\right)^2\right)\right)a_m^{(2)}a_n^{(2)}\right) \\
&+\frac{1}{2J}\left(\bar{I}_1-3\right)\left(2\eta_1\bar{B}_{mi}D_{ij}\bar{B}_{jn}+2\left(\eta_2\bar{J}_4-\frac{1}{3}\eta_1 tr(d)\right)\bar{B}_{mk}\bar{B}_{nk}\right. \\
&\left.-\frac{1}{3}\bar{J}_4\left(\eta_1+2\eta_2\left(\bar{I}_1^2-2\bar{I}_2\right)\right)\delta_{mn}\right)
\end{aligned}
\tag{8.5}
$$

$$
\begin{aligned}
\mathrm{DDSDDE} &= \left(\frac{4}{3}\left(C_{10}\bar{I}_1+4C_{01}\bar{I}_2\right)+\kappa\left(2J-1\right)\right)\delta_{mn}\delta_{rs} \\
&+\left(2\bar{I}_1C_{10}+4\bar{I}_2C_{01}+\kappa\left(1-J\right)\right)\left(\delta_{mr}\delta_{ns}+\delta_{nr}\delta_{ms}\right) \\
&-4\left(C_{10}+2C_{01}\bar{I}_1\right)\left(\bar{B}_{mn}\delta_{rs}+\bar{B}_{rs}\delta_{mn}\right) \\
&+4C_{01}\left(2\left(\bar{B}_{rk}\bar{B}_{sk}\delta_{mn}+\bar{B}_{mk}\bar{B}_{nk}\delta_{rs}\right)-2\bar{B}_{mn}\bar{B}_{rs}+\left(\bar{B}_{mr}\bar{B}_{ns}+\bar{B}_{nr}\bar{B}_{ms}\right)\right) \\
&-\frac{1}{2}\varepsilon\left(\left(\delta_{mn}\delta_{rs}-\left(\delta_{mr}\delta_{ns}+\delta_{ms}\delta_{nr}\right)\right)\mathrm{e}_l\mathrm{e}_l\right) \\
&+\varepsilon\left(\mathrm{e}_n\mathrm{e}_m\delta_{rs}+\mathrm{e}_r\mathrm{e}_s\delta_{mn}-\mathrm{e}_n\mathrm{e}_s\delta_{mr}-\mathrm{e}_m\mathrm{e}_r\delta_{ns}-\mathrm{e}_m\mathrm{e}_s\delta_{nk}-\mathrm{e}_n\mathrm{e}_r\delta_{ms}\right) \\
&+\frac{4}{J}\left(k_1^{(1)}\left(1+2k_2^{(1)}\left(I_4-1\right)^2\right)\exp\left(k_2^{(1)}\left(I_4-1\right)^2\right)a_m^{(1)}a_n^{(1)}a_r^{(1)}a_s^{(1)}\right. \\
&\left.+k_1^{(2)}\left(1+2k_2^{(2)}\left(I_6-1\right)^2\right)\exp\left(k_2^{(2)}\left(I_6-1\right)^2\right)a_m^{(2)}a_n^{(2)}a_r^{(2)}a_s^{(2)}\right) \\
&+\frac{1}{J}\left(\bar{I}_1-3\right)\left(\frac{1}{2}\eta_1\left(\bar{B}_{mr}\bar{B}_{ns}+\bar{B}_{ms}\bar{B}_{nr}\right)+2\eta_2\bar{B}_{mk}\bar{B}_{nk}\bar{B}_{rl}\bar{B}_{sl}\right. \\
&-\frac{1}{3}J^{-\frac{4}{3}}\left(\eta_1+2\eta_2\left(\bar{I}_1^2-2\bar{I}_2\right)\right)\left(\bar{B}_{mk}\bar{B}_{nk}\delta_{rs}+\bar{B}_{rk}\bar{B}_{sk}\delta_{mn}\right) \\
&\left.+\frac{1}{9}\left(\eta_1\left(\bar{I}_1^2-2\bar{I}_2\right)+2\eta_2\left(\bar{I}_1^2-2\bar{I}_2\right)^2\right)\delta_{mn}\delta_{rs}\right)
\end{aligned}
\tag{8.6}
$$

To cover the viscoelastic effect, the Cauchy stress would change to:

$$\sigma^{n+1} = \text{DDSDDE}\epsilon^{n} + \sum_{j=1}^{N'} \exp\left(-\frac{\Delta t}{\tau_j}\right) h_j^n + \gamma_j \tau_j \frac{\left(1-\exp\left(-\frac{\Delta t}{\tau_j}\right)\right)}{\Delta t}\left(\text{DDSDDE}\epsilon^{n+1} - \text{DDSDDE}\epsilon^{n}\right) \tag{8.7}$$

Consequently, the iterative formula for the consistent tangent modulus would be:

$$\text{DDSDDE}^{n+1} = \frac{\partial \Delta\sigma}{\partial \Delta\epsilon} = \left\{1 + \sum_{j=1}^{3} \gamma_j \tau_j \frac{\left(1-\exp\left(-\frac{\Delta t}{\tau_j}\right)\right)}{\Delta t}\right\} \text{DDSDDE}^{n} \tag{8.8}$$

We implemented the model proposed in Chapter 7 to capture the behavior of a multi-layered actuator. According to the model:

$$\boldsymbol{\sigma} = \sum_{q=1}^{\mathbb{N}} v_{(q)} \boldsymbol{\sigma}_{(q)} \tag{8.9}$$

where $v_{(q)}$ and $\boldsymbol{\sigma}_{(q)}$ are the volume fraction and strain energy of the $q$th layer, respectively. $\boldsymbol{\sigma}_{(q)}$ is replaced by the expressions (8.2) or (8.7) for the $q$th layer.

## 8.3 MODEL CALIBRATION AND VALIDATION

To calibrate the model, we use the experimental results reported in previous studies. Then the experimental results are used to validate the model. Since there is not experimental analysis on a fiber-reinforced dielectric elastomer actuator considering all the mentioned effects, we validate fiber-reinforced hyperelastic, dielectric, rate-dependent, viscoelastic and multi-layered behavior separately.

### 8.3.1 Model Calibration for Hyperelastic and Anisotropic Constants

In experimental tests, the results are typically described using the nominal stress. This stress is obtained by dividing the applied force by the cross-sectional area in the reference coordinates. For an incompressible material under simple tension, the nominal stress, $\mathcal{T}$, is expressed as follows:

$$\mathcal{T} = \frac{dW}{d\lambda} \tag{8.10}$$

The strain energy function is denoted as $W\left(\lambda, \lambda^{-1/2}\right)$. This energy function depends on the principal stretches: $\lambda_1 = \lambda$ and $\lambda_2 = \lambda_3 = \lambda^{-1/2}$. To calibrate and validate the hyperelastic and anisotropic components of our model, we extracted experimental data from a simple uniaxial tensile test [1]. The deformation gradient tensor and the right Cauchy-Green strain tensor, denoted as (**C**), were determined for this loading condition.

$$\boldsymbol{F} = \begin{bmatrix} \lambda & 0 & 0 \\ 0 & \lambda^{-1/2} & 0 \\ 0 & 0 & \lambda^{-1/2} \end{bmatrix} \tag{8.11}$$

$$\boldsymbol{C} = \begin{bmatrix} \lambda^2 & 0 & 0 \\ 0 & \lambda^{-1} & 0 \\ 0 & 0 & \lambda^{-1} \end{bmatrix} \tag{8.12}$$

Therefore, the Jacobian first and second invariants are:

$$J = 1 \tag{8.13}$$

$$I_1 = \lambda^2 + 2\lambda^{-1} \tag{8.14}$$

$$I_2 = \lambda^{-2} + 2\lambda \tag{8.15}$$

The hyperelastic part of the strain energy function in the model is Presented using the Mooney-Rivlin model.

$$W^{\text{hyperelastic}} = C_{10}\left(\lambda^2 + 2\lambda^{-1} - 3\right) + C_{01}\left(\lambda^{-2} + 2\lambda - 3\right) \tag{8.16}$$

To determine the anisotropic constants of the energy function, unit vectors, and corresponding invariants along the fiber direction in the reference coordinates must be computed. The selected study reported the experimental tests involved samples reinforced by two families of fibers at angles $\theta$ and $-\theta$. The fiber arrangement is depicted in Figure 8.1.

So, we have:

$$\boldsymbol{A}^{(1)} = \begin{Bmatrix} \cos\alpha \\ \sin\alpha \\ 0 \end{Bmatrix} \tag{8.17}$$

$$\boldsymbol{A}^{(2)} = \begin{Bmatrix} \cos(-\alpha) \\ \sin(-\alpha) \\ 0 \end{Bmatrix} = \begin{Bmatrix} \cos\alpha \\ -\sin\alpha \\ 0 \end{Bmatrix} \tag{8.18}$$

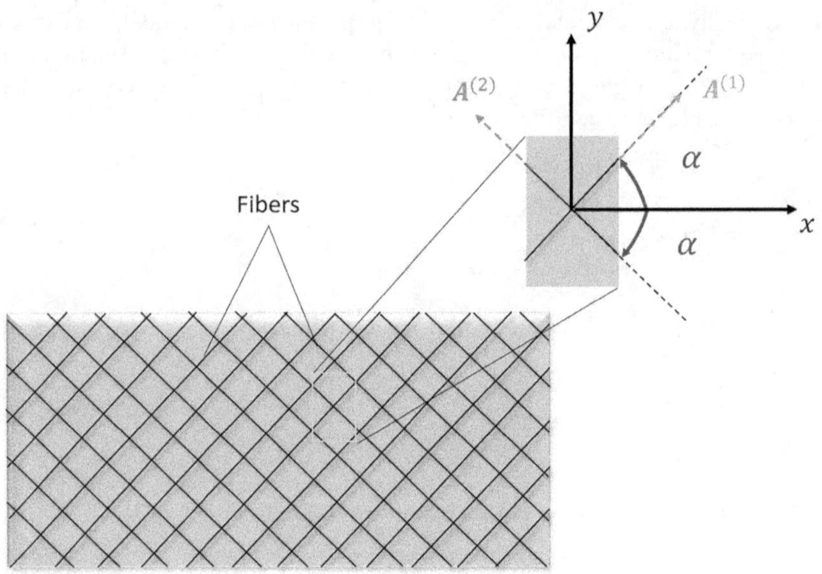


**FIGURE 8.1** Fibers configuration in an element.

$$I_4 = \boldsymbol{A}^{(1)}.\boldsymbol{C}\boldsymbol{A}^{(1)} = \begin{Bmatrix} \cos\alpha \\ \sin\alpha \\ 0 \end{Bmatrix} . \begin{bmatrix} \lambda^2 & 0 & 0 \\ 0 & \lambda^{-1} & 0 \\ 0 & 0 & \lambda^{-1} \end{bmatrix} \begin{Bmatrix} \cos\alpha \\ \sin\alpha \\ 0 \end{Bmatrix} = \lambda^2\cos^2\alpha + \lambda^{-1}\sin^2\alpha \tag{8.19}$$

$$I_6 = \boldsymbol{A}^{(2)}.\boldsymbol{C}\boldsymbol{A}^{(2)} = \begin{Bmatrix} \cos\alpha \\ -\sin\alpha \\ 0 \end{Bmatrix} . \begin{bmatrix} \lambda^2 & 0 & 0 \\ 0 & \lambda^{-1} & 0 \\ 0 & 0 & \lambda^{-1} \end{bmatrix} \begin{Bmatrix} \cos\alpha \\ -\sin\alpha \\ 0 \end{Bmatrix} = \lambda^2\cos^2\alpha + \lambda^{-1}\sin^2\alpha \tag{8.20}$$

Using the calculated parameters, $I_4$ and $I_6$, the strain energy function for the anisotropic section is obtained as Equation (8.21).

$$W^{\text{anisotropic}} = \frac{k_1}{k_2}\left\{\exp\left[k_2\left(\lambda^2\cos^2\alpha + \lambda^{-1}\sin^2\alpha - 1\right)^2\right] - 1\right\} \tag{8.21}$$

Once the expression of the strain energy function is completed, $\mathcal{T}$, the nominal stress is extracted according to Equation (8.10):

$$\begin{aligned} \mathcal{T} = \frac{dW}{d\lambda} &= \frac{dW^{\text{hyperelastic}}}{d\lambda} + \frac{dW^{\text{anisotropic}}}{d\lambda} = 2\left(C_{10}\lambda + C_{01}\right)\left(1-\lambda^{-3}\right) \\ &+2k_1\left(2\lambda^3\cos^4\alpha - \lambda^{-3}\sin^4\alpha + \cos^2\alpha\sin^2\alpha + \lambda^{-2}\sin^2\alpha - 2\lambda\cos^2\alpha\right) \\ &\left\{\exp\left[k_2\left(\lambda^2\cos^2\alpha + \lambda^{-1}\sin^2\alpha - 1\right)^2\right]\right\} \end{aligned} \tag{8.22}$$

**TABLE 8.1**
**Hyperelastic and Anisotropic Constants for Cotton Fiber Embedded in the Silicon Rubber Matrix**

| $C_{10}$ (Mpa) | $C_{01}$ (Mpa) | $\kappa$(Mpa) | Fiber Angle | $\kappa_1$(Mpa) | $k_2$ |
|---|---|---|---|---|---|
| 0.12780 | 0.00139 | 25.838 | 0° | 78.93 | 0.1 |
| | | | ±15° | 12.08 | |
| | | | ±30° | 3.07 | |
| | | | ±45° | 1.88 | |

To calibrate the nominal stress, a curve-fitting procedure is applied using the available experimental results in [1] and the derived equation. Hyperelastic and anisotropic constants for cotton fiber embedded in the silicon rubber matrix (in different fiber angles) are obtained as listed in Table 8.1.

Figure 8.2 shows the experimental results compared to Equation (8.22) and the obtained constants.

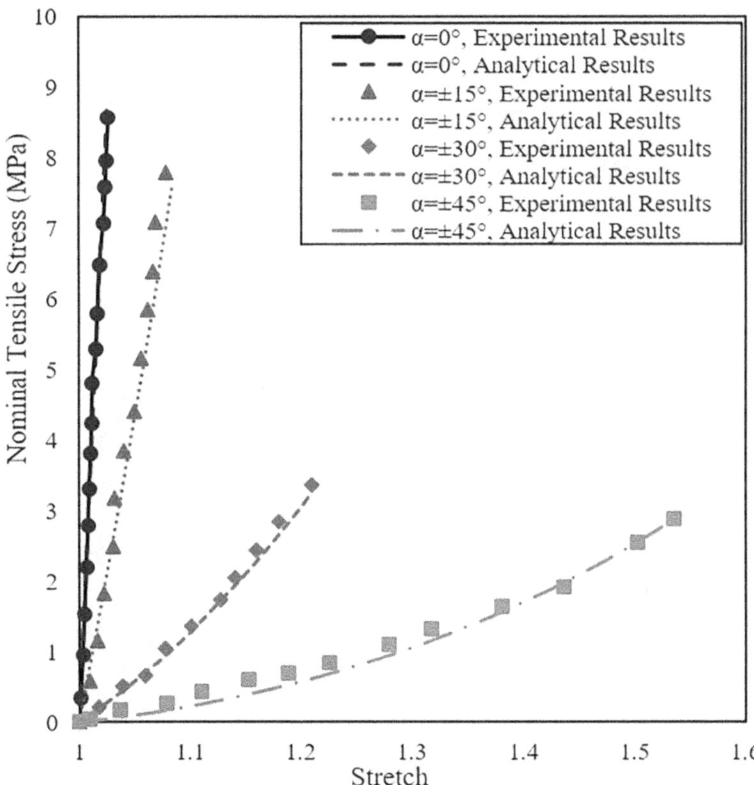


**FIGURE 8.2** Experimental [1] and analytical results for anisotropic hyperelastic material.

### 8.3.2 Model Validation

To assess the effectiveness of the proposed model and its capability to address specific characteristics, some Finite Element Method (FEM) analyses using Abaqus were conducted. The simulations implement the UMAT subroutine developed based on the equations discussed earlier.

First, experimental tensile tests were simulated to validate the model performance regarding hyperelastic and anisotropic properties separately. Then, the simulation of a uniaxial compression test shows the model's ability to predict strain rate dependency caused by the viscoelastic characteristics of the elastomer. Finally, comparing experimental data with FEM-obtained results for a DE actuation ensures the adequacy of the derived model to take the dielectric effect into account.

Therefore, a FEM analysis to simulate a uniaxial tensile test was conducted. The simulation results were used to validate the hyperelastic terms of the model. Inspired by the reference experimental test, the simulated parts was a 50 mm × 50 mm × 1 mm patch under tensile test [2]. The properties of VHB 4910 ($C_{10} = 0.016(\text{Mpa})$ and $C_{01} = 0.0073(\text{Mpa})$ [3]). Figure 8.3 presents the experimental setup and stress contour of FEM analysis.

Figure 8.4 compares the stress-strain results of the experiment and numerical analysis. According to the acceptable differences between the results, model captures the hyperelastic behavior accurately enough.

Next, a FEM model of a fiber-reinforced dog bone specimen under tensile test was simulated using Abaqus, similar to the experimental tensile test specimen reported in [1]. Geometrical features and fiber configuration are displayed in Figure 8.5.

Twelve thousand, eight-node reduced-integration linear brick elements (C3D8R) were employed to discretize the dog bone specimen. Figure 8.6 represents the obtained results of the simulation in comparison with experimental test results. In addition to a good agreement between the results, stiffer behavior in specimens

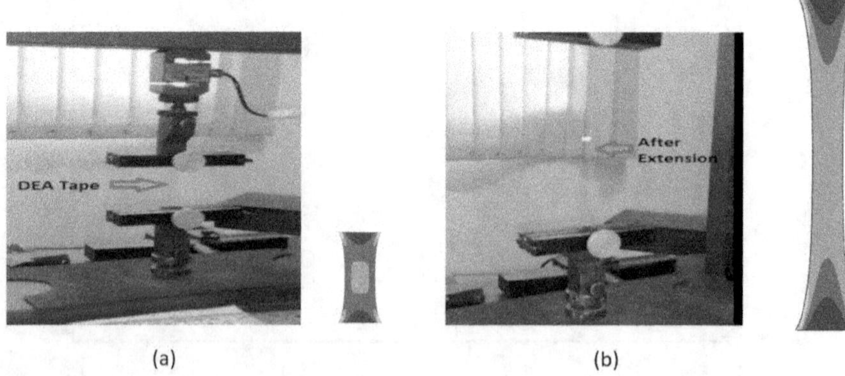


**FIGURE 8.3** Uniaxial tensile test of hyperelastic material, experiment setup [2], and FEM stress contour (a) Before extension; (b) After extension.

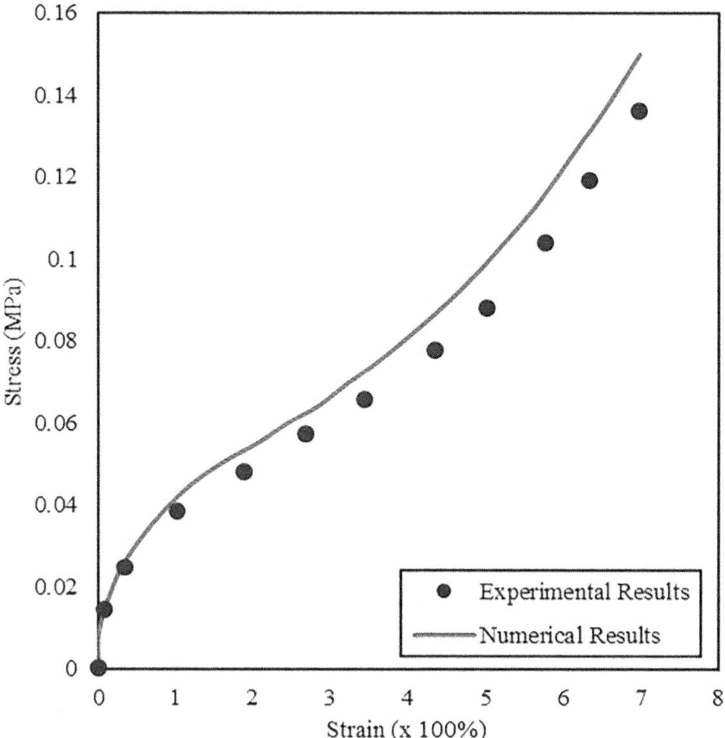


**FIGURE 8.4** Experimental [2] and analytical results for hyperelastic material.

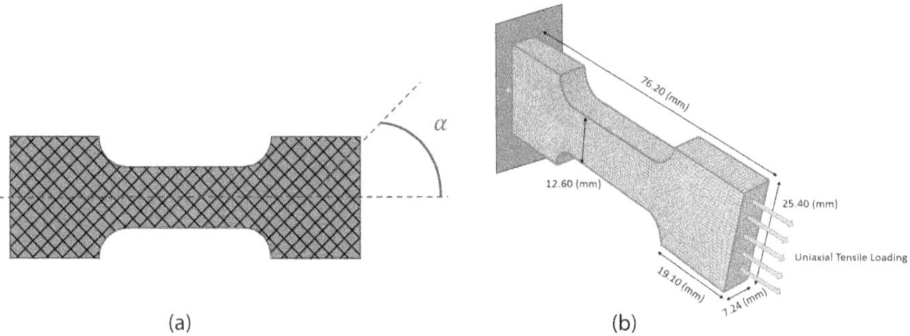


**FIGURE 8.5** Specimen for tensile test. (a) Fiber configuration embedded in the specimen; (b) Specimen geometry, boundary conditions, and loading.

with smaller fiber angles is observed. More details on the validation process are presented in [4.6].

In the next step, an unconfined uniaxial compression test of a cylindrical sample was simulated, inspired by an experimental test in [7]. Geometrical details are presented in Figure 8.7. The lower base of the sample is constrained in the vertical

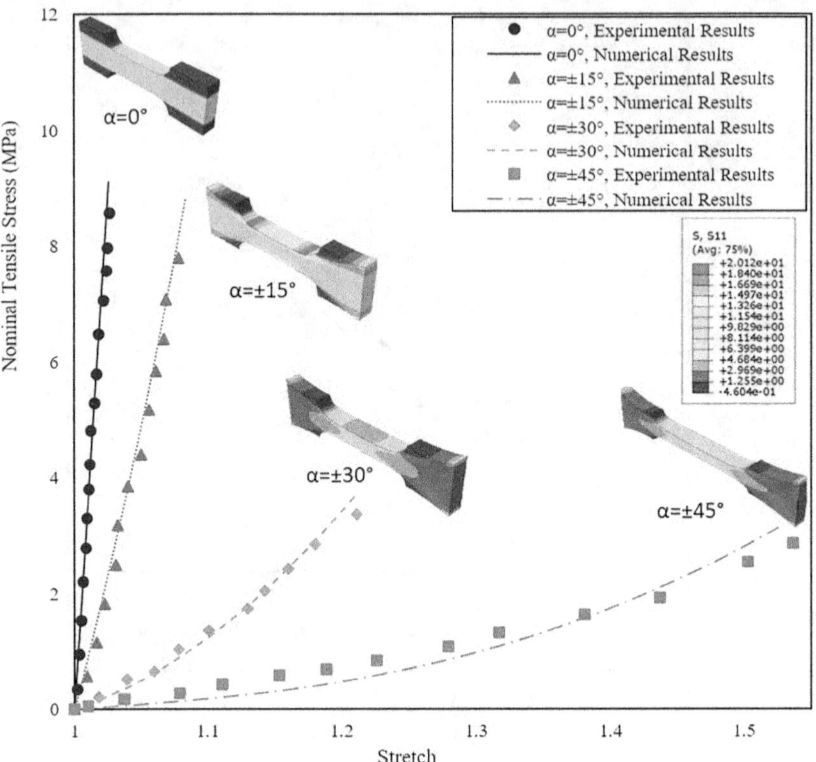


**FIGURE 8.6** (a) Experimental [1] and numerical tensile test results for specimens with different fiber angles. (b) Uniaxial tensile test of anisotropic hyperelastic material at different fiber orientations, FEM stress contour.

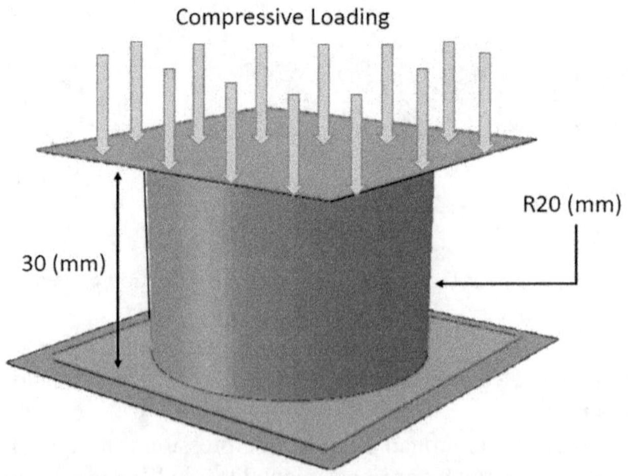


**FIGURE 8.7** Cylindrical configuration for compressive test.

**TABLE 8.2**
**Hyperelastic and Viscoelastic Constants [7]**

| $C_{10}$ (kpa) | $C_{01}$ (kpa) | $\kappa$(kpa) | $\eta_1$(kpa · s) | $\eta_2$(kpa · s) |
|---|---|---|---|---|
| −0.484 | 0.785 | 800 | 670 | 300 |

direction while a 3 mm displacement (in the form of two different velocities, 0.176 mm/s and 0.0216 mm/s) is applied on the other base along the cylinder axis. Symmetric geometry and loading make it possible to reduce the model to an axisymmetric model and use 150 four-node bilinear axisymmetric quadrilateral elements (CAX4).

The reported material parameters in [7] for a hyper-viscoelastic liver tissue sample were used for the analysis and are mentioned in Table 8.2.

In addition to modeling and numerical analysis, the relationship between nominal stress and strain can be obtained and depicted using analytical equations. For the desired loading, $\boldsymbol{F}$ and $\boldsymbol{C}$ are similar to the expressions obtained in the previous section. However, in this section, the time derivative of matrix $\boldsymbol{C}$ is needed.

$$\dot{\boldsymbol{C}} = \begin{bmatrix} 2\lambda\dot{\lambda} & 0 & 0 \\ 0 & -\lambda^{-2}\dot{\lambda} & 0 \\ 0 & 0 & -\lambda^{-2}\dot{\lambda} \end{bmatrix} \tag{8.23}$$

Also, the invariants of $\dot{C}$ are

$$J_2 = \left(2\lambda^2 + \lambda^{-4}\right)\dot{\lambda}^2 \tag{8.24}$$

$$J_4 = 2\left(\lambda^3 - \lambda^{-3}\right)\dot{\lambda} \tag{8.25}$$

By using these parameters in the relationship selected for the viscoelastic section, the strain energy function in this section is obtained as follows:

$$W^{\text{viscoelastic}} = \frac{1}{4}\left(\lambda^2 + 2\lambda^{-1} - 3\right)\left[\eta_1\left(2\lambda^2 + \lambda^{-4}\right) + \eta_2\left(2\left(\lambda^3 - \lambda^{-3}\right)\right)^2\right]\dot{\lambda}^2 \tag{8.26}$$

Next, we calculate the derivative of this term with respect to $\dot{\lambda}$, which is a part of the nominal stress in this loading.

$$\frac{dW^{\text{viscoelastic}}}{d\dot{\lambda}} = \frac{1}{2}\left(\lambda^2 + 2\lambda^{-1} - 3\right)\left[\eta_1\left(2\lambda^2 + \lambda^{-4}\right) + \eta_2\left(2\left(\lambda^3 - \lambda^{-3}\right)\right)^2\right]\dot{\lambda} \tag{8.27}$$

Thus, the final nominal stress $\mathcal{T}$ is:

$$\mathcal{T} = \frac{dW^{\text{hyperelastic}}}{d\lambda} + \frac{dW^{\text{viscoelastic}}}{d\dot{\lambda}} = 2\left(C_{10}\lambda + C_{01}\right)\left(1 - \lambda^{-3}\right)$$
$$+\frac{1}{2}\left(\lambda^2 + 2\lambda^{-1} - 3\right)\left[\eta_1\left(2\lambda^2 + \lambda^{-4}\right) + \eta_2\left(2\left(\lambda^3 - \lambda^{-3}\right)\right)^2\right]\dot{\lambda} \quad (8.28)$$

Using the coefficients presented in Table 8.2 in this equation will lead to the results that are compared with the results of the experimental test shown in Figure 8.8 and Figure 8.9. Strain rate dependency is evident in the presented results. The figures also show the model accuracy in following experimental data, especially in lower rate loading.

The fourth step of validation is dedicated to the DE effect. In this step, inspired by an experiment reported in Ref. [8], the actuation of a DE film is simulated. The simulation includes pre-stretches along two perpendicular directions, too. The material parameters are considered for a VHB 4905 film as $C_{10} = 0.016$ (Mpa), $C_{01} = 0.016$ (Mpa) [3], and $\varepsilon_r = 4.16$ [9]. The model is discretized by 1200 elements.

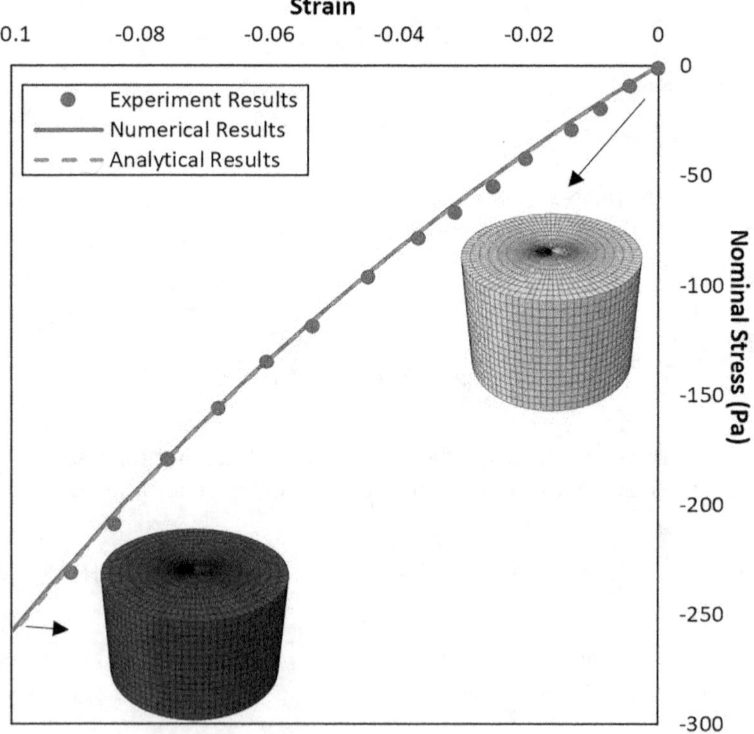


**FIGURE 8.8** Stress-Strain curve (experimental [7], numerical, analytical result), loading rate 0.0216 mm/s.

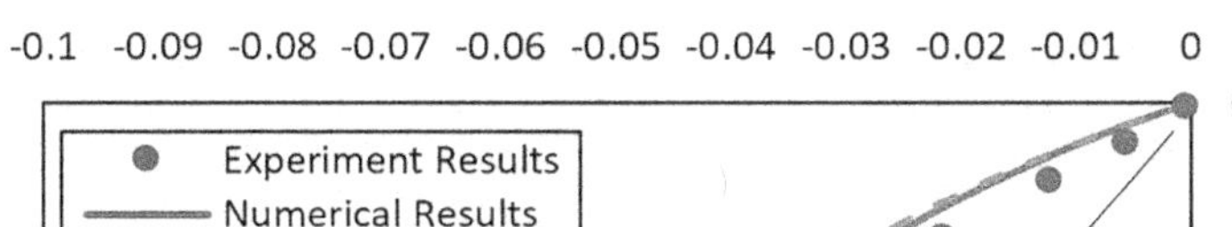


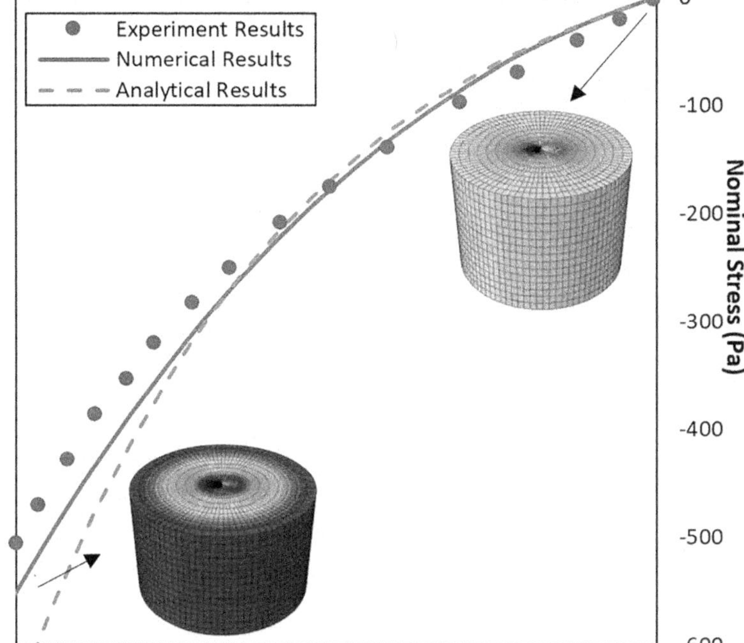


**FIGURE 8.9** Stress-Strain curve (experimental [7], numerical, analytical result), loading rate 0.176 mm/s.

Mechanical pre-stretches are followed by actuation along the film thickness caused by a 5kV voltage. Since the proposed model needs an electric field, using Equation (8.29), the nominal electric field is calculated. In this equation, $L_3$ is the initial thickness of the film (before pre-stretches).

$$E_3 = \frac{\mathbb{V}}{L_3} \tag{8.29}$$

Figure 8.10 shows the test setup, schematic illustration and simulation of loading, and boundary conditions during the actuation step.

On the other hand, the analytical results can be compared with those obtained from the experimental test and numerical analysis. In the present case, the elastomer made of VHB 4905 is first subjected to mechanical load $P_1$ and changes from the reference state (undeformed) to the pre-stretched state. In this step, applying of mechanical load will cause stretch $\lambda_{P1}$ in the direction of the load. Then an electric field will be applied to the sample and reach the actuated state. At this

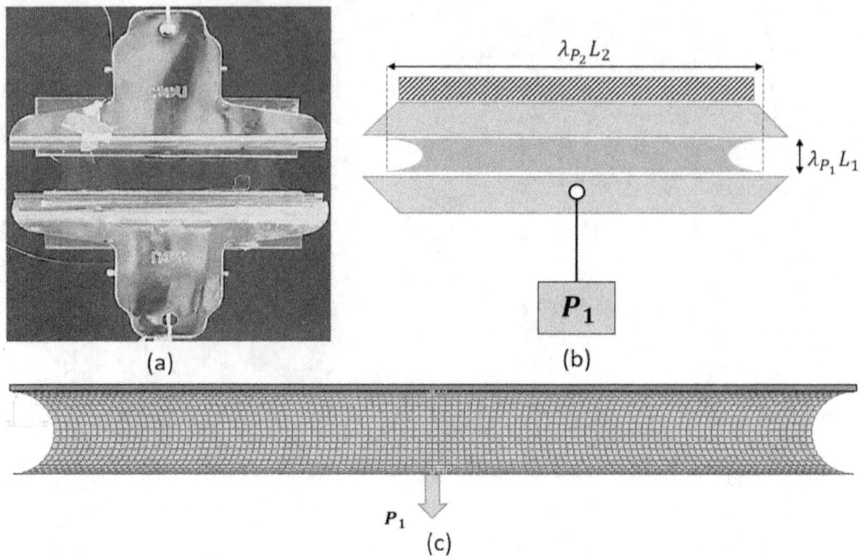


FIGURE 8.10 Pre-stretch state. (a) Experimental setup [8]; (b) Schematic illustration of loading and boundary condition; (c) Loading and boundary condition simulation.

stage, the stretch in the in-plane directions is $\lambda_1$ and $\lambda_2$ compared to the reference state (undeformed). Also, according to the assumption of incompressibility of the sample, the stretch in the direction perpendicular to the plane will be $\lambda_3 = \dfrac{1}{\lambda_1 \lambda_2}$. The loading steps are shown in Figure 8.11.

In this loading, the equation of state is expressed as follows:

$$\sigma_1 = \lambda_1 \frac{\partial W^{\text{hyperelastic}}}{\partial \lambda_1} + \lambda_1 \frac{\partial W^{\text{electric}}}{\partial \lambda_1} \tag{8.30}$$

where $\sigma_1$ is the stress caused by the application of mechanical load.

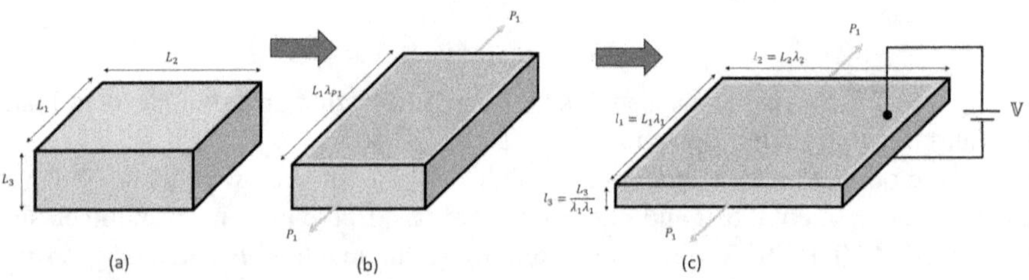


FIGURE 8.11 Mechanical and electrical loading of hyper dielectric sample. (a) Reference state; (b) Pre-stretch state; (c) Actuated state.

Based on the deformations created in the sample and applied electrical stimulation, the deformation gradient tensor, electric field vector, hyperelastic and dielectric parts of the energy function are:

$$\boldsymbol{F} = \begin{bmatrix} \lambda_1 & 0 & 0 \\ 0 & \lambda_2 & 0 \\ 0 & 0 & \lambda_1^{-1}\lambda_2^{-1} \end{bmatrix} \tag{8.31}$$

$$\mathbb{E} = \begin{bmatrix} 0 \\ 0 \\ E_3 \end{bmatrix} \tag{8.32}$$

$$W^{\text{hyperelastic}} = C_{10}\left(\lambda_1^2 + \lambda_2^2 + \lambda_1^{-2}\lambda_2^{-2} - 3\right) + C_{01}\left(\lambda_1^{-2} + \lambda_2^{-2} + \lambda_1^2\lambda_2^2 - 3\right) \tag{8.33}$$

$$W^{\text{electric}} = -\frac{1}{2}\varepsilon J\boldsymbol{C}^{-1} : \mathbb{E}^2 = -\frac{1}{2}\varepsilon\lambda_1^2\lambda_2^2 E_3^2 = -\frac{1}{2}\varepsilon\lambda_1^2\lambda_2^2\left(\frac{\Phi}{L_3}\right)^2 \tag{8.34}$$

Finally, considering that the value of $\lambda_{P1}$ in the pre-stretch state (before applying the voltage) is reported to be 2.36, the stretch-voltage relation can be shown in Figure 8.12. The results confirm the model's possibility of predicting the DE effect on the mechanical behavior for strains less than 200%. This constraint is caused by the Mooney-Rivlin model limitation in predicting hyperplastic behavior in strains larger than 200%.

To validate the proposed model for a multi-layered dielectric elastomer, a comprehensive comparison between the analytical vs. numerical and numerical vs. experimental results for hyperelastic base multi-layered structures was conducted.

Firstly, the analytical Piola-Kirchhoff stress for a two- and three-layer fiber-reinforced dielectric sample under a uniaxial tensile test was computed. This approach was inspired by Ref. [2], previously used to validate the single-layer model. In this case, the dog bone sample was assumed to be made up of layers with half and one-third of the thickness of the final sample. The layers were considered to be reinforced with two families of fibers at ±30° and ±45°. The schematic representation of the multi-layered samples is shown in Figure 8.13.

The stress of each fiber-reinforced single layer is expressed as:

$$\begin{aligned} T_\eta = \frac{dW_\eta}{d\lambda} = \frac{dW_\eta^{\text{iso}}}{d\lambda} + \frac{dW_\eta^{\text{aniso}}}{d\lambda} = 2\left(C_{10}\lambda + C_{01}\right)\left(1 - \lambda^{-3}\right) \\ +2k_1\left(2\lambda^3\cos^4\alpha - \lambda^{-3}\sin^4\alpha + \cos^2\alpha\sin^2\alpha + \lambda^{-2}\sin^2\alpha - 2\lambda\cos^2\alpha\right) \\ \left\{\exp\left[k_2\left(\lambda^2\cos^2\alpha + \lambda^{-1}\sin^2\alpha - 1\right)^2\right]\right\} \end{aligned} \tag{8.35}$$

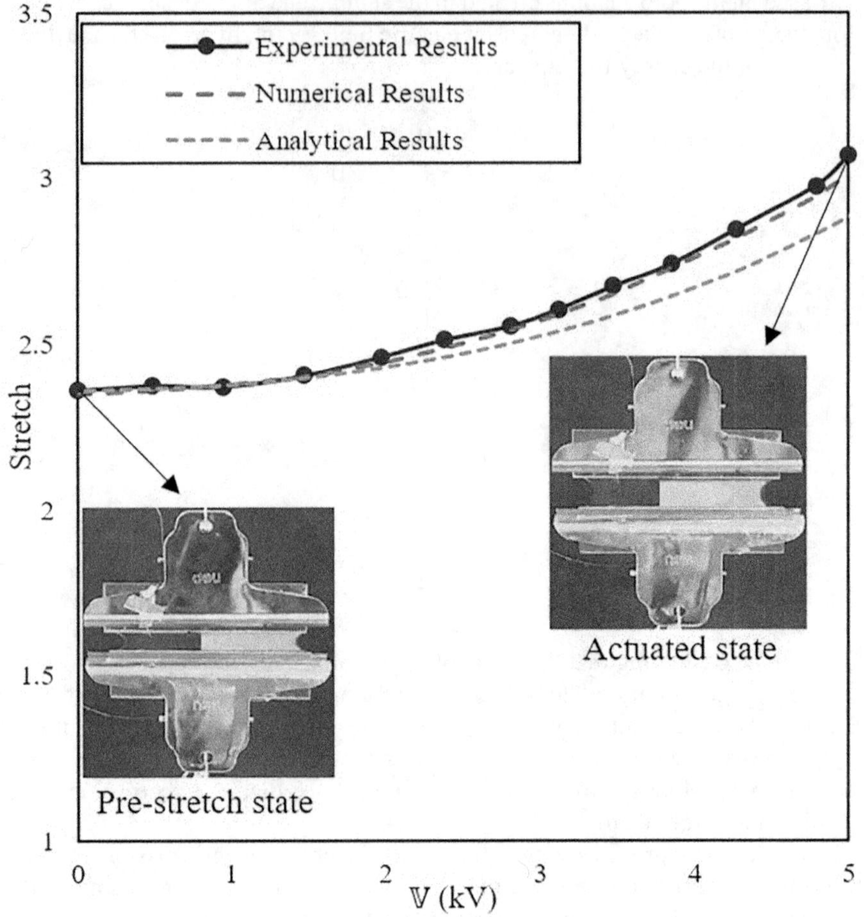

**FIGURE 8.12** Stretch-voltage chart for elastomer dielectric material under electrical and mechanical loading (experimental [8], numerical, analytical result).

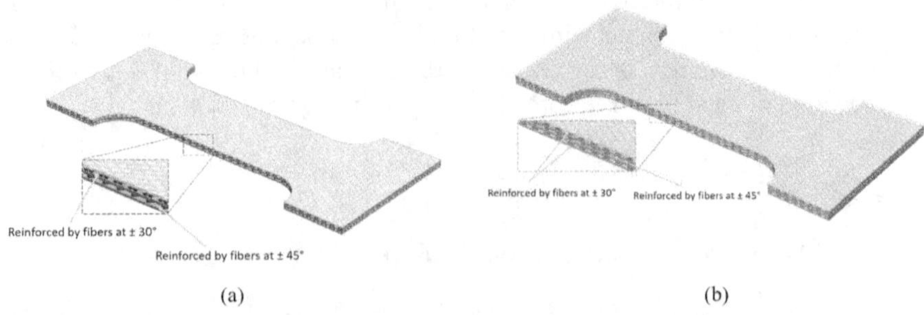


**FIGURE 8.13** Simulated (a) 2-layer and (b) 3-layer dog bone samples.

For layers with two fiber families at ±30° and ±45°, the stress is:

$$\begin{aligned} T_{\pm 30} &= 2\left(C_{10}\lambda + C_{01}\right)\left(1-\lambda^{-3}\right) \\ &+2k_1\left(2\lambda^3\cos^4\alpha - \lambda^{-3}\sin^4\alpha + \cos^2\alpha\sin^2\alpha + \lambda^{-2}\sin^2\alpha - 2\lambda\cos^2\alpha\right) \\ &\left\{\exp\left[k_2\left(\lambda^2\cos^2\alpha + \lambda^{-1}\sin^2\alpha - 1\right)^2\right]\right\} \end{aligned} \tag{8.36}$$

$$\begin{aligned} T_{\pm 45} &= 2\left(C_{10}\lambda + C_{01}\right)\left(1-\lambda^{-3}\right) \\ &+2k_1\left(2\lambda^3\cos^4\alpha - \lambda^{-3}\sin^4\alpha + \cos^2\alpha\sin^2\alpha + \lambda^{-2}\sin^2\alpha - 2\lambda\cos^2\alpha\right) \\ &\left\{\exp\left[k_2\left(\lambda^2\cos^2\alpha + \lambda^{-1}\sin^2\alpha - 1\right)^2\right]\right\} \end{aligned} \tag{8.37}$$

Assuming equal thickness (and volume) for each layer, which is half of the whole structure, the stress of the bilayer is:

$$T_{2\text{-layer}} = v_{\pm 30}T_{\pm 30} + v_{\pm 45}T_{\pm 45} = \frac{1}{2}\left(T_{\pm 30} + T_{\pm 45}\right) \tag{8.38}$$

A similar geometrical model was simulated in Abaqus and analyzed using the developed UMAT code.

A 3-layer fiber-reinforced hyperelastic dog bone was also simulated. For the 3-layer model, each layer's thickness was 33% of the whole thickness, while the intermediate layer was 34% thick. The outer and intermediate layers were reinforced with fiber families at ±30° and ±45°, respectively. The stress of this multi-layered structure is determined accordingly:

$$\begin{aligned} T_{3\text{-layer}} &= v_{\pm 30}T_{\pm 30} + v_{\pm 45}T_{\pm 45} + v_{\pm 30}T_{\pm 30} = 0.33\times T_{\pm 30} \\ &+ 0.34\times T_{\pm 45} + 0.33\times T_{\pm 30} \end{aligned} \tag{8.39}$$

Figure 8.14 depicts analytical vs. numerical nominal stress for 2-layer and 3-layer samples under uniaxial tests. The results are reasonably close.

On the other hand, to assess the model's performance using experimental results, we simulated and studied the binder reported in Ref. [10]. The binder comprises two, three, and four active silicon layers constrained with a stiff steel substrate. The substrate is 50 mm long and 11 mm wide (Figure 8.15), while the active layers' dimensions are 90 × 15 × 0.05 mm. An electric field is applied to the active layers. The corresponding voltage ranges from 0 to about 3.5 kV. Figure 8.15 illustrates a simulated and experimental sample of a multi-layered dielectric elastomer actuator.

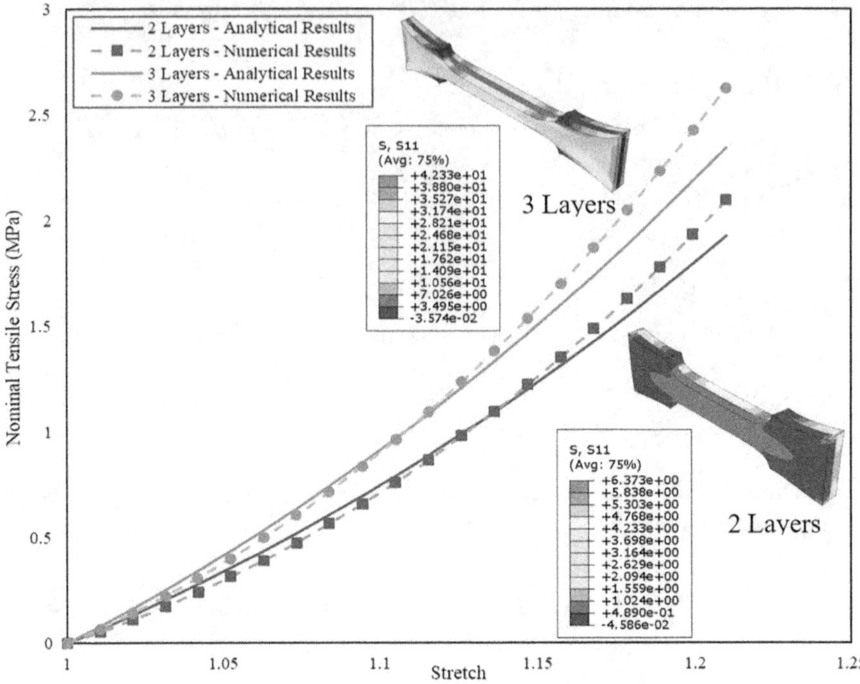


**FIGURE 8.14** Analytical and numerical results for the uniaxial tensile test of multi-layered samples.

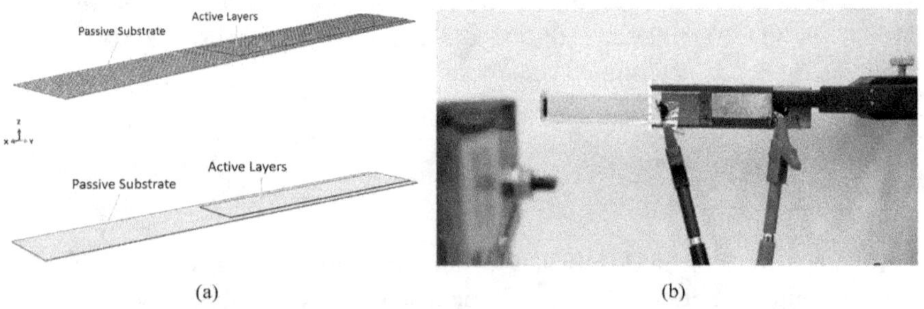


**FIGURE 8.15** (a) Simulated and (b) Experimental [10] multi-layered dielectric elastomer actuator.

The deflection of the binder's tip for growing voltage levels was documented experimentally and extracted from the numerical solution. The results are depicted in Figure 8.16 and exhibit reasonable agreement for the dielectric multi-layered actuator. Details about the multilayer model validation are available in [11].

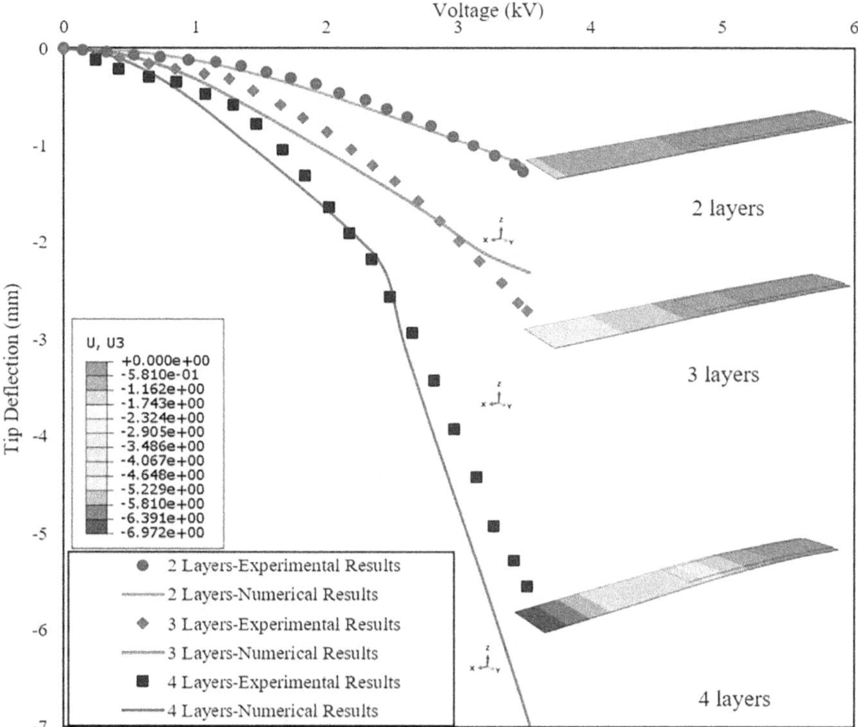


**FIGURE 8.16** Numerical and experimental [10] tip deflection results for multi-layered dielectric elastomer.

## 8.4 NUMERICAL IMPLEMENTATION OF PRACTICAL CASE STUDIES

The models are assumed to be VHB 1049 patches reinforced with one or two families of cotton fibers. The hyperelastic, anisotropic, and dielectric constants for the material are extracted using curve-fitting processes or available data, which are listed in Table 8.3.

For anisotropic analyses, in addition to hyperelastic properties, anisotropic constants are needed, which are presented in Table 8.3. For the model, the effective dielectric constant for the final composite, $\varepsilon_{\text{eff}}$, is calculated using relation (8.40), in which $\varepsilon_m, \varepsilon_f$, and $v_{(m)}$ stand for the dielectric constant of the matrix, the dielectric constant of inclusion, and the volume fraction of the inclusions, respectively.

$$\left(\frac{\varepsilon_{\text{eff}} - \varepsilon_m}{\varepsilon_{\text{eff}} + 2\varepsilon_m}\right) = v_{(m)}\left(\frac{\varepsilon_f - \varepsilon_m}{\varepsilon_f + 2\varepsilon_m}\right) \tag{8.40}$$

**TABLE 8.3**
**Hyperelastic, Anisotropic and Dielectric Constants for Numerical Analysis**

| Isotropic Constants | | Anisotropic Constants | | | Relative Permittivity | | |
|---|---|---|---|---|---|---|---|
| $C_{10}$ (MPa) | $C_{01}$ (MPa) | $\alpha$ | $k_1$(MPa) | $k_2$ | $\varepsilon_{rm}$ | $\varepsilon_{rf}$ | $\varepsilon_{reff}$ |
| −0.012 | 0.039 | 0° | 78.93 | 0.1 | 4.7 | 3.18 | 3.87 |
| | | ±15° | 12.08 | | | | |
| | | ±30° | 3.07 | | | | |
| | | ±45° | 1.88 | | | | |

Available references have reported the relative permittivity of cotton to be 3.18 [12]. The fiber volume fraction is assumed to be 51.8%. Therefore, the effective dielectric constant for the anisotropic layer would be 3.87.

### 8.4.1 Bilayer Anisotropic Bending Dielectric Elastomer Actuator

The section is aimed to propose a structure consisting of an isotropic and an anisotropic layer. The anisotropic layer is coated by thin electrode layers on both sides to provide electroactive behavior. This structure can be used in various geometrical features under different loading and boundary conditions. Other versions of the structure make actuators with unique configurations and applications. Figure 8.17 depicts this structure schematically.

According to the extracted relations for Cauchy stress and tangent modulus and the combination of the discussed features, it is possible to develop software

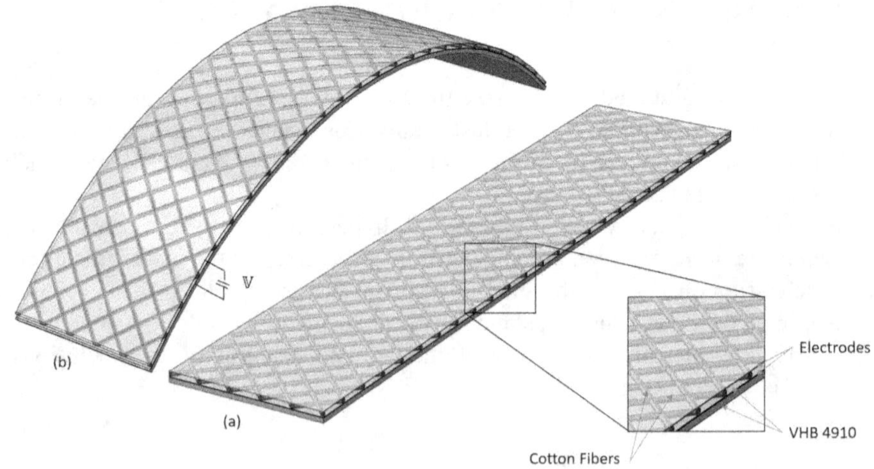


**FIGURE 8.17** Schematic of an anisotropic DE bending actuator. (a) Reference state (b) Actuated state.

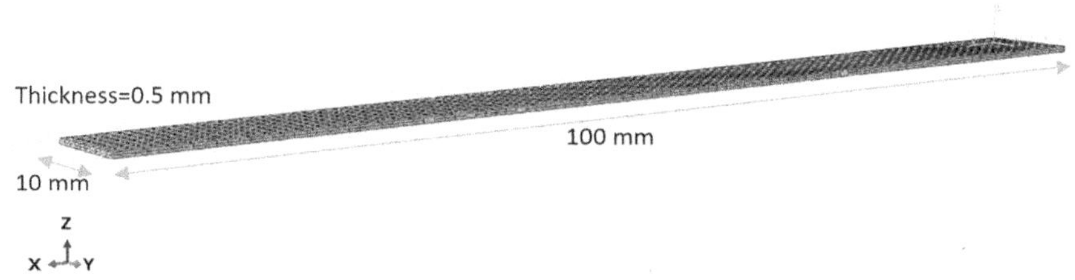


**FIGURE 8.18** Dimensions and discretization of bending specimen designed for numerical analysis.

codes for materials with three different behaviors, including hyper viscoelastic material, dielectric elastomer material, and dielectric elastomer, considering rate-dependent viscoelastic properties. In each analysis, based on the purpose of the investigation, the appropriate code and properties can be used. Once the UMAT subroutine is completed, to use it and study the effect of each parameter on the stress changes of the material, a beam was modeled. Half of the thickness of the beam is assumed to be electrically inactive and isotropic, made of VHB 4910. The other half is electroactive (VHB 4910 between two electrodes), anisotropic (cotton fibers embedded in VHB 4910 matrix).

The modeled beam is fixed at one end, and the boundary conditions at the other end are selected according to the purpose of the study. 2000 C3D8R elements have also been used to discretize the model. Details are presented in Figure 8.18.

In one end, all the displacements are constrained, while the other end is entirely free to deform. An electric field of 12 (Mvolt/m) was applied to the plate in the direction of its thickness.

Displacement contour for anisotropic plates, presented in Figure 8.19, reveals the significant influence of fiber direction on plate deformation. According to the results, coupled bending and twisting behavior is evident in anisotropic plates. Coupled behavior makes anisotropic DE material a reasonable candidate for producing various actuators. While the distance of the final profile to $y = 0$ (vertical displacement) presents bending deformation, the final shape of the profile depicts twisting behavior. Figure 8.19 confirms that each fiber direction dictates unique behavior that is useful for specific purposes.

In the first anisotropic analysis, the simulated model is actuated by an electric field in the direction of actuator thickness. It can be seen that the amount of blocking force decreases with the addition of fibers (Figure 8.20). But contrary to expectations, the force of the beam with fibers at ±45 is slightly higher than the force of the beam with fibers at 0.

According to Figure 8.21, addition of fibers in any direction reduces the deflection in the beam subjected to the electric field.

Next, the anisotropic beam with the mentioned specifications is subjected to a mechanical load (velocity of 20 mm/s at the free end of the beam). It can be seen in Figure 8.22 that the addition of fibers with a zero-degree angle causes a

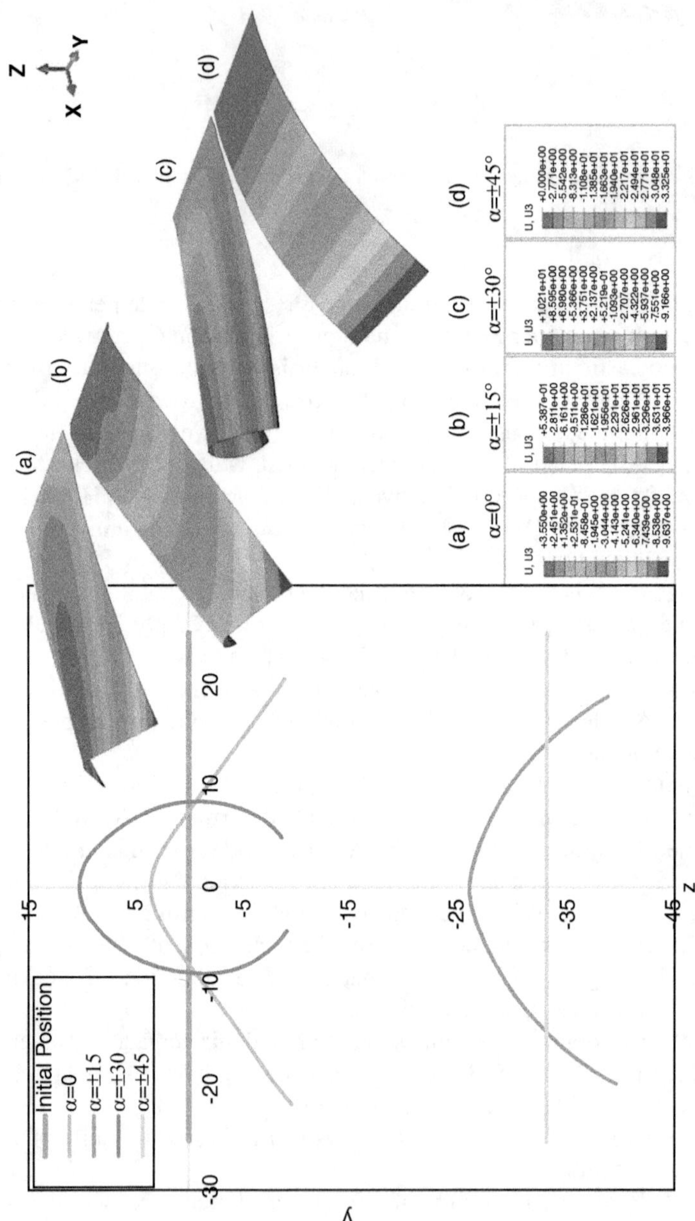


**FIGURE 8.19** Free end profile of anisotropic dielectric elastomer with different fiber orientations.

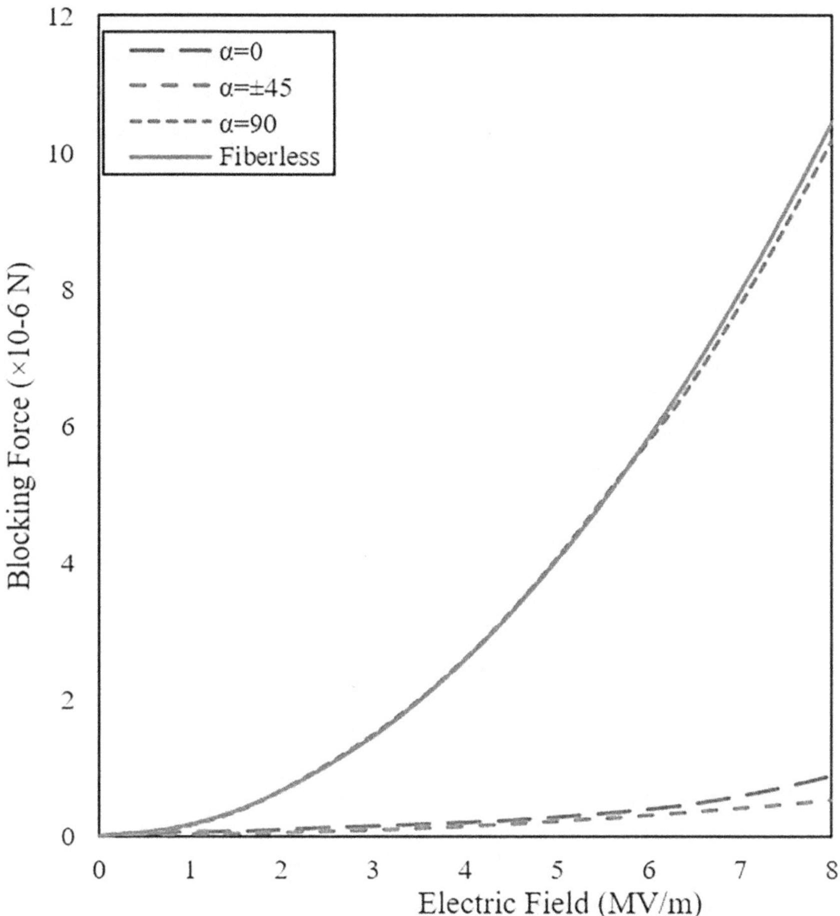


**FIGURE 8.20** Blocking force for anisotropic DE actuators actuated by the electrical load.

significant increase in the reaction force. The increase in reaction force due to the reinforcements, with a minor difference, is also observed in other angles.

By comparing the force-displacement results of an anisotropic beam under electric loading and rate-dependent mechanical loading in Figure 8.23, it seems that the dielectric property (especially in large fiber angles and isotropic beams) leads to nonlinearity. In contrast, the behavior of the visco beam under velocity at the free end is more similar to linear behavior.

Considering the viscoelastic and dielectric properties simultaneously, the behavior of the beam subjected to an electric field and velocity of 20 mm/s at the free end of the beam is studied. In this condition, it can be seen that the reaction force in the beam with 90-degree fibers is slightly higher than the isotropic beam. An increase in reaction force can be seen more clearly by increasing the fiber angles for material reinforced by fibers at ±45 and 90.

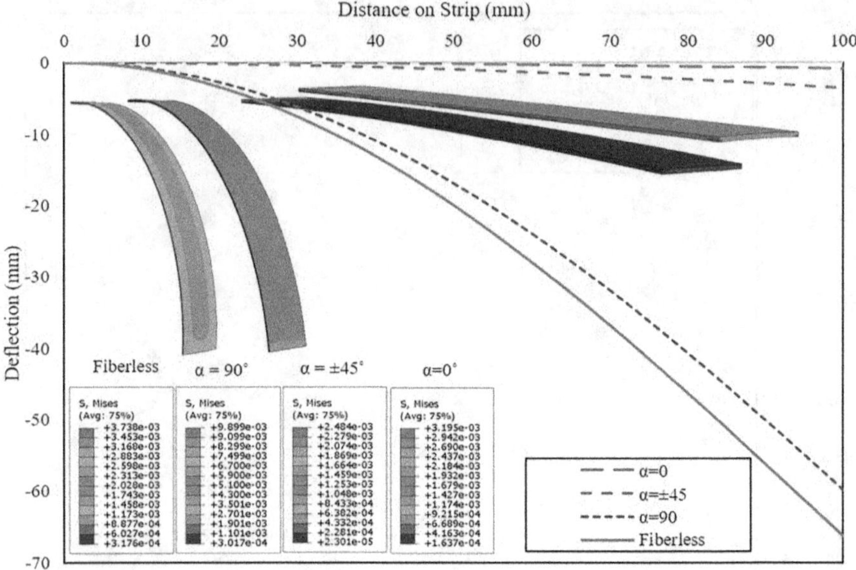


FIGURE 8.21 Vertical deformation for anisotropic DE actuators actuated by the electrical load.

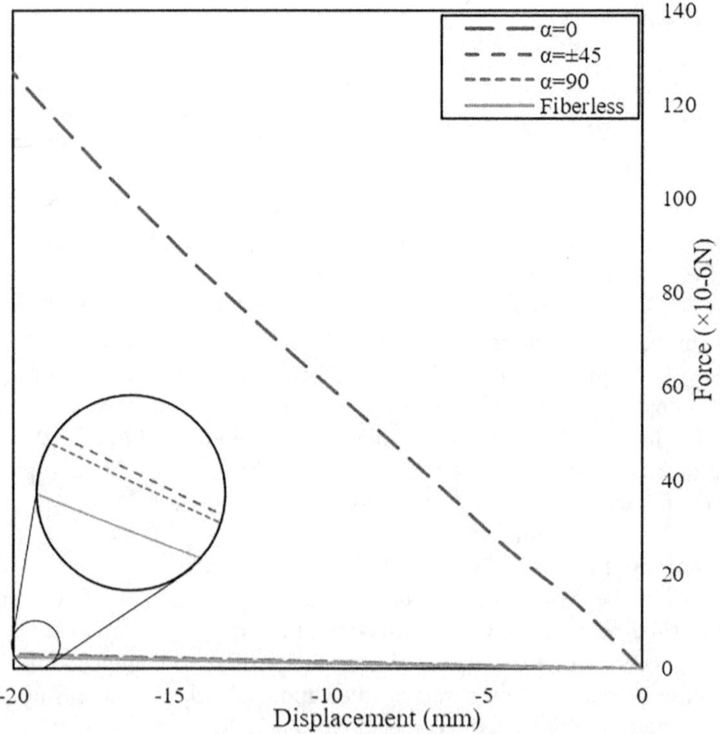


FIGURE 8.22 Reaction force for anisotropic DE actuator under mechanical load.

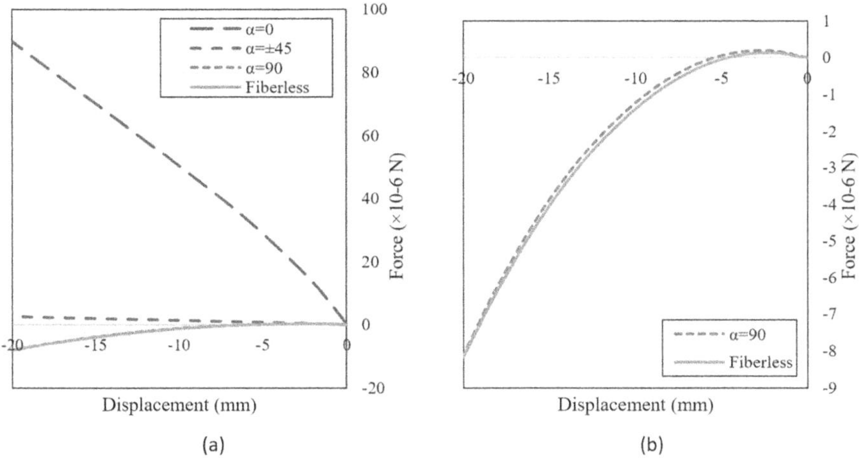


**FIGURE 8.23** Reaction force for (a) Anisotropic DE actuators under mechanical and electrical load (b) Results of the DE actuators with selected fiber direction DE actuators.

It is noteworthy that the amount of reaction force for isotropic and anisotropic beams with fibers at 90 degrees, which was positive at small deformations and the beginning of loading, gradually decreased and reached a negative value with the increase of beam deflection. This parameter, however, is always positive for an anisotropic beam with ±45- and 0-degree fibers.

The beam bending diagram (Figure 8.24) also shows that the deflection of the anisotropic beam under electrical and mechanical load in fibers with a smaller

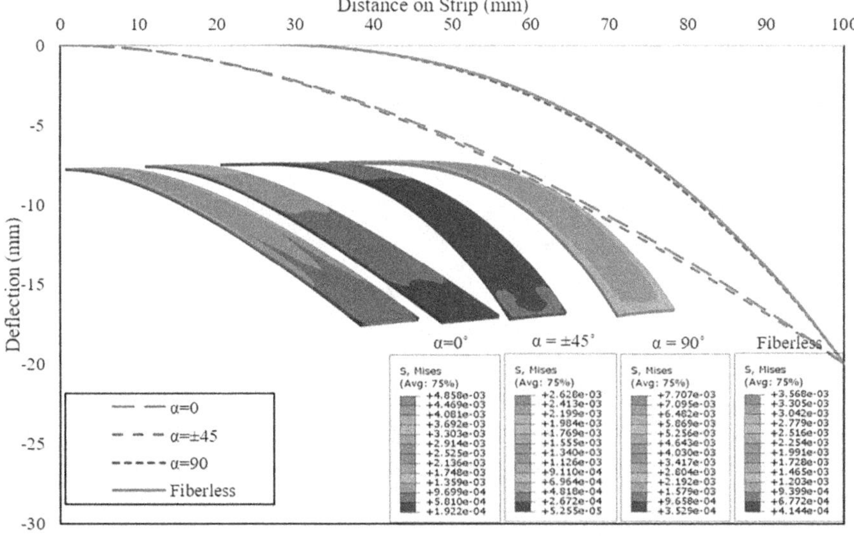


**FIGURE 8.24** Vertical deformation for anisotropic DE actuators under mechanical and electrical load.

angle is closer to linear behavior. In small angles of fibers, the visco effect has a more decisive role in how the shape of the beam changes. However, in larger angles and isotropic beams, the dielectric property has a more significant impact on the material's behavior, and for this reason, nonlinearity is seen in these beams. In an isotropic beam, the reason why the dielectric property prevails over the visco property is the sizeable relative permittivity of the material.

To study rate dependency for an anisotropic beam, an analysis similar to the visco isotropic beam was carried out for the DE bending actuator reinforced by fibers at different angles (an electric field of 8 mV/m is applied). The applied velocity for beams with different fiber angles was chosen based on the maximum possible amount for complete analysis. Therefore, the final deflections for the beams are different. According to Figure 8.25, rate-dependent behavior could be seen in an anisotropic actuator. Results of the bending actuator loaded by pure mechanical loading are presented in Ref. [13].

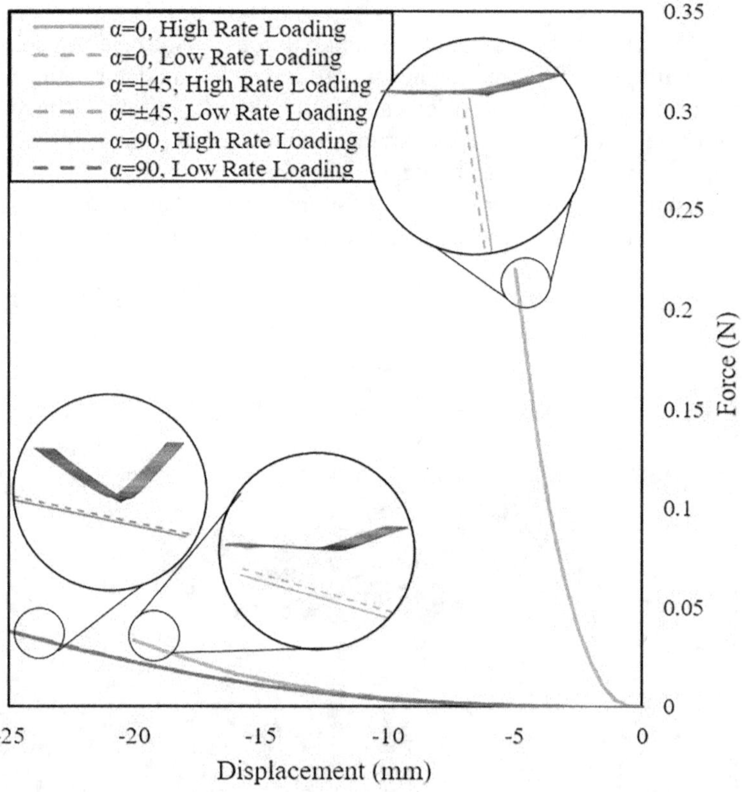


**FIGURE 8.25** Rate-dependent behavior of an anisotropic DE beam under mechanical and electrical load.

### 8.4.2 Single-Layer Tubular Dielectric Elastomer Actuator

To study the numerical results of a tubular actuator, we simulated an actuator made of an elastomer, reinforced by a family of fibers, and coated on both sides with thin and soft electrodes. The dimensions are $L = 100$ (mm), $R = 10$ (mm), and $H = 2$ (mm). The actuator's bottom base is assumed to be entirely fixed, while the other is free. Furthermore, an electric field is applied to deform the actuator.

The model to be examined using Abaqus is simulated with a discretization of 13,600, 8-node linear brick elements. Depending on the purpose of the analysis, external mechanical loads such as internal pressure, axial force, or torque would be applied. Figure 8.26 illustrates the geometric and loading features of the simulated tubular dielectric elastomer actuator (TDEA).

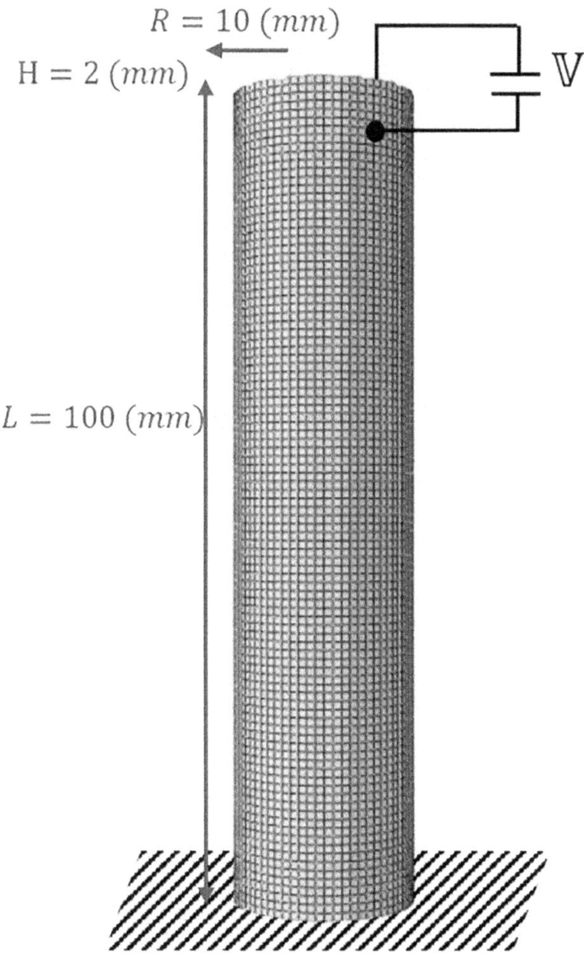


**FIGURE 8.26** Geometrical and Loading Features of the Simulated TDEA.

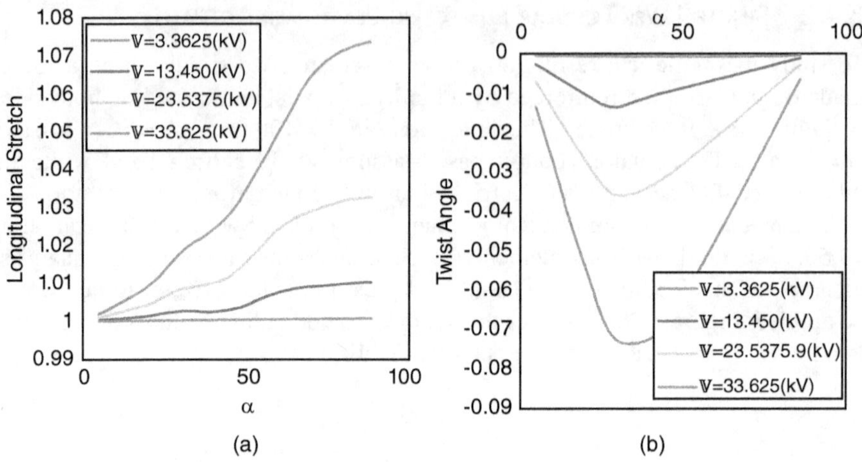


**FIGURE 8.27** (a) Longitudinal stretch - Fiber angle and (b) Twist angle - Fiber angle for low voltage analytical analyses.

#### 8.4.2.1 Low Voltage Analysis

Figure 8.27 presents the changes in longitudinal and twist angles with fiber orientation. Both longitudinal stretch and twist angle grow with activating voltage. Additionally, it is detectable that as the voltage rises, the sensitivity of these parameters to fiber orientation increases. A larger fiber angle leads to a longer final length; nonetheless, for a twist angle, there is an optimum fiber angle around $\alpha = 35°$, for the most considerable twist angle. It is worth noting that in the TDEA, fibers with orientations of 0° and 90° provide the smallest twist angle due to their axisymmetric structure. The readers are encouraged to see Ref. [14] for semi-analytical analysis of anisotropic tubular actuator and results.

According to Figure 8.28, for $\alpha = 90°$, numerical analyses show a hoop stretch of less than one. According to numerical results, the horizontal fiber reinforcement of the DE actuator results in a more significant longitudinal stretch than the fiberless actuator. Smaller fiber angles, on the other hand, have the opposite impact on longitudinal deformation. However, regardless of the orientation of the reinforcement, fiber reinforcing reduces hoop stretch.

Figure 8.29 displays vertical displacement contours and twisting angles for both fiberless and fiber-reinforced actuators subjected to voltage. The $z$-axis deformation is scaled by a factor of 3 for clarity.

#### 8.4.2.2 Effects of Internal Pressure

To investigate more details about a fiber-reinforced DE actuator, further numerical analyses were conducted using a radial internal pressure, $P_r$, of 1.5 (kPa). Electrical loading was applied at three voltage levels: 16 (kV), 24 (kV), and 32 (kV). The exact boundary conditions and properties were used as in the previous analyses.

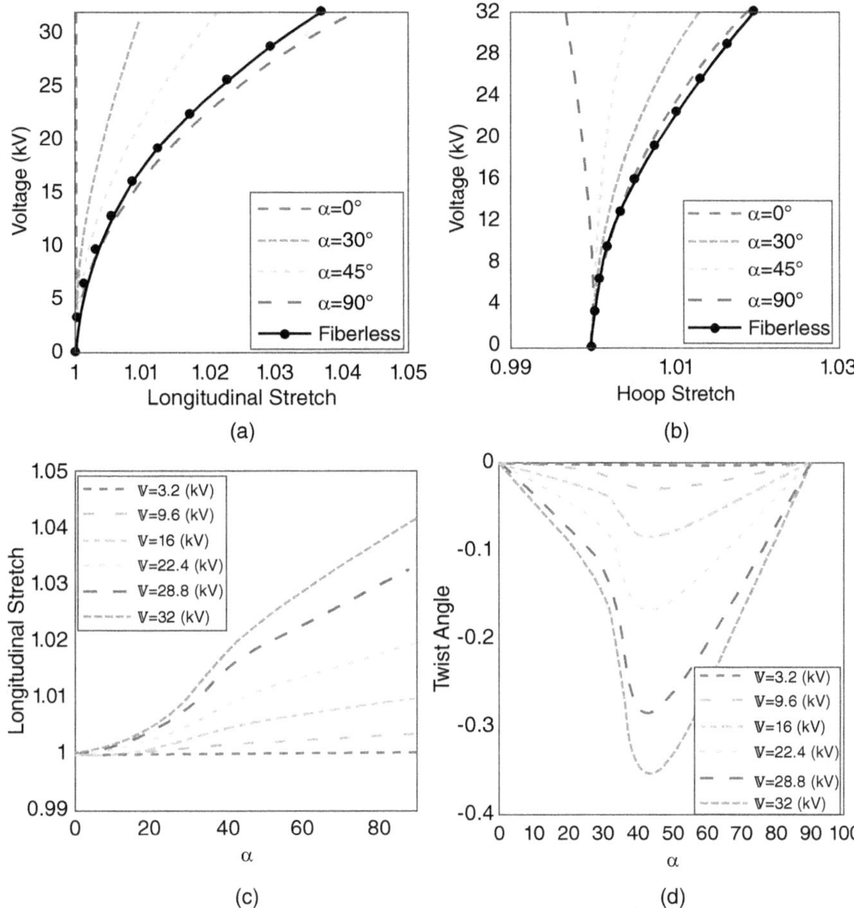


**FIGURE 8.28** Numerical results for low voltage analyses (a) Voltage-longitudinal stretch (b) Voltage-hoop stretch (c) Longitudinal stretch-fiber angle (d) Twist angle-fiber angle.

Figure 8.30 illustrates the longitudinal stretch in the presence of internal pressure and an electric field. Both figures confirm the eventual increase in the longitudinal stretch because of higher voltages. For horizontal and vertical fibers, higher voltages do not change the trend of the results. However, for fiber angles between 15 and 45 degrees, the trend of the curve changes, and the slope decreases initially. In other words, as the voltage increases, its impact on the longitudinal deformation (tendency to lengthening) grows and resists the effect of internal pressure (tendency to shortening).

Figure 8.30(b) depicts that the internal pressure plays a role in the longitudinal deformation of actuators reinforced by fibers at $\alpha = 15°$, $\alpha = 30°$, and $\alpha = 45°$. At the lowest voltage level, longitudinal deformation grows (longitudinal stretch goes down) as α increases. However, for $\mathbb{V} = 24 \text{ and } 32\left(\text{kV}\right)$, the rise in α first

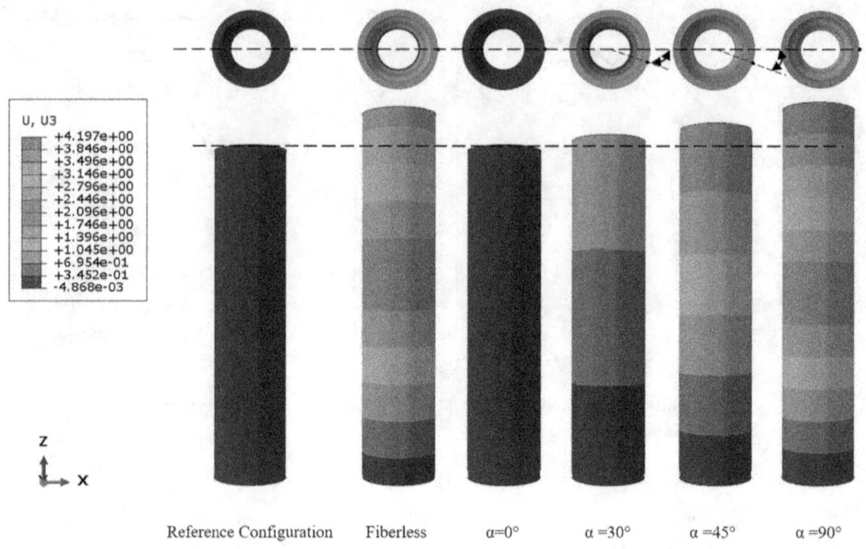


**FIGURE 8.29** Vertical displacement contours and twisting angle of numerical analyses for electrical actuation, $\mathbb{V} = 32\left(\mathrm{Kv}\right)$ (longitudinal deformation scale factor = 3).

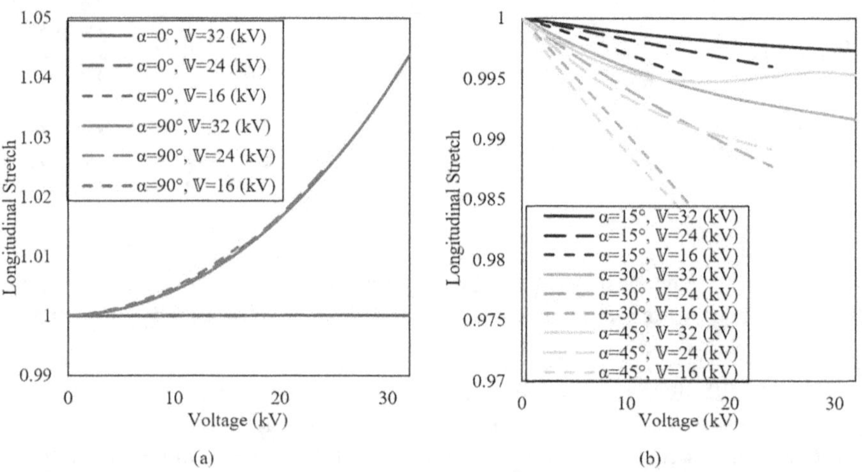


**FIGURE 8.30** Numerical results of longitudinal stretch-voltage under internal pressure (a) $\alpha = 0°$, and $\alpha = 90°$, (b) $\alpha = 15°$, $\alpha = 30°$, and $\alpha = 45°$.

decreases the longitudinal stretch and then levels it up. At $\mathbb{V} = 16\left(\mathrm{kV}\right)$, the internal pressure shortens the actuator; however, for higher voltages, it loses its power to determine the deformation direction for $30° \leq \alpha \leq 45°$.

Figure 8.31 shows that actuators with fibers at $\alpha = 0°$ and $\alpha = 90°$ display the smallest twist angle. For $\alpha = 45°$, the twist angle is always negative. Its amount decreases as the applied electric load rises. For $\alpha = 30°$, the twist angle is positive;

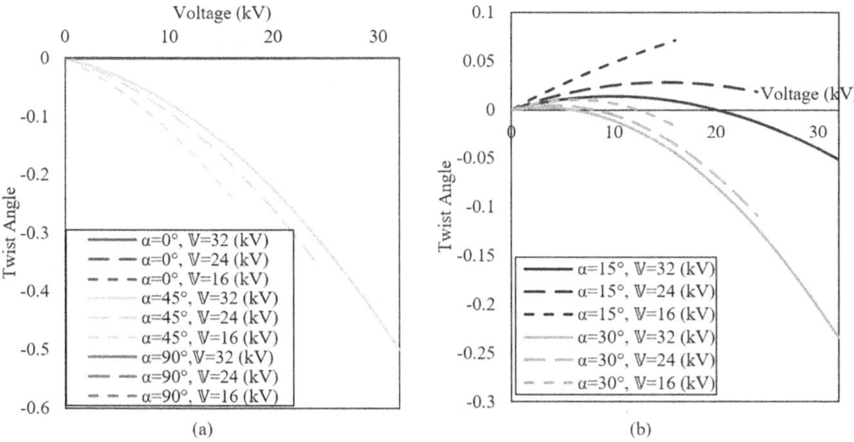


**FIGURE 8.31** Numerical results of twist angle - voltage, under internal pressure (a) $\alpha$ = 0°, $\alpha$ = 45°, and $\alpha$ = 90°, (b) $\alpha$ = 15° and $\alpha$ = 30°.

however, its slope decreases with voltage growth. The actuator with fibers at $\alpha = 15°$ at the beginning of the analysis produces positive twist angles and a positive slope in the twist angle curve. The higher the final electric load, the sooner the positive phase ends. A greater final twist angle is observed for the higher final electric load.

The results indicate that an axisymmetric arrangement of the fibers results in no twist angle, whereas the maximum amount of twist angle is observed at $\alpha = 45°$ for the fibers. It is also detectable that the electric field induces a negative twist angle, and at the range of $0° < \alpha < 45°$, as α grows, the electric field dominates the twist angle more.

Figure 8.32 illustrates the hoop stretch of the actuator under applied voltages. Referring to Figure 8.32, in a given voltage, a lower electric load drives more considerable hoop stretches since as the electric load decreases, internal pressure can cause hoop stretch. However, this trend is not valid for the final hoop stretch. As the voltage rises, the final hoop stretches for actuators with fibers at $\alpha$ = 90° declines, while it grows for fibers at $\alpha$ = 45° and $\alpha$ = 0°. According to Figure 8.32(b), hoop stretches curves for $\alpha = 15°$, and $\alpha = 30°$ change their positions with each other as the electric load increases.

Figure 8.33 displays the vertical displacement contours and twisting angle for fiberless and fiber-reinforced actuators subjected to voltage and internal pressure.

#### 8.4.2.3 The Behavior of TDEA Under Axial Load

To investigate the effect of constant axial load on the behavior of an electrically actuated fiber-reinforced TDEA, some numerical analyses were conducted. Results are presented for three different forces: $F_z = 0$, $F_z = 0.6(\text{N})$, and $F_z = 1.2(\text{N})$. Figure 8.34 displays the results of actuators reinforced by different fiber orientations and fiberless actuators.

According to Figure 8.34(a), the combination of the applied voltage and axial load cannot cause a longitudinal stretch in the actuator with fibers at $\alpha$ = 0°.

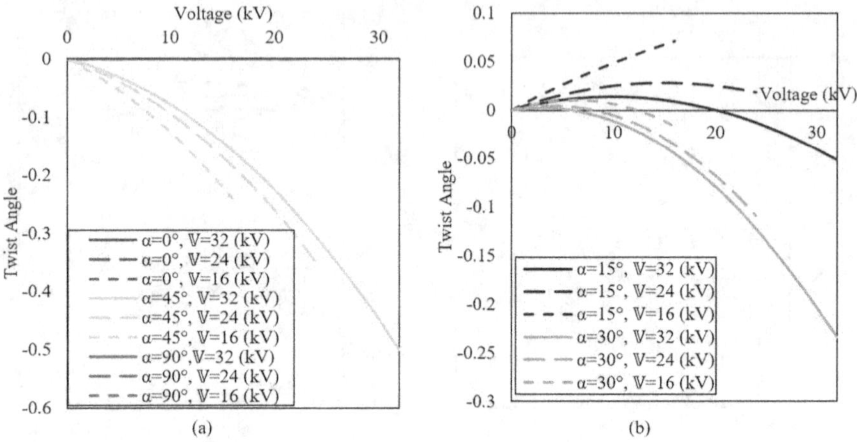


**FIGURE 8.32** Numerical results of hoop stretch - voltage, under internal pressure (a) $\alpha = 0°$, $\alpha = 45°$and $\alpha = 90°$, (b) $\alpha = 15°$ and $\alpha = 30°$.

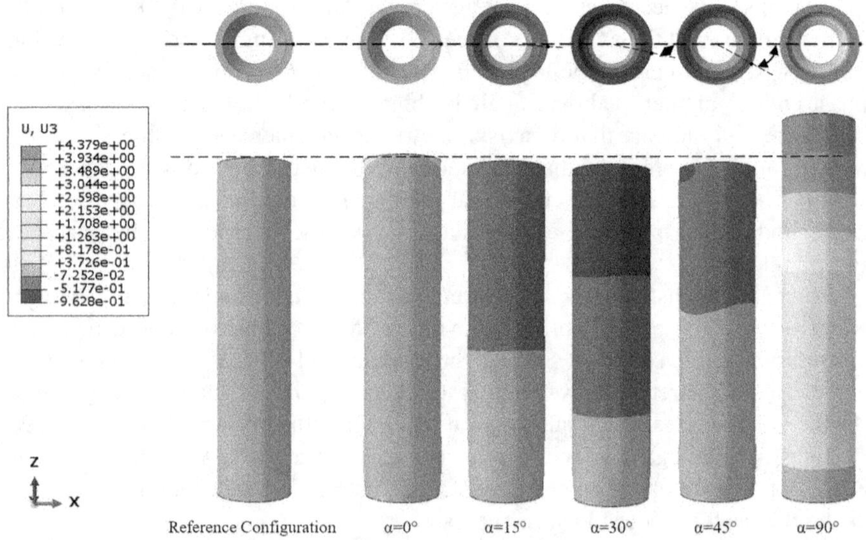


**FIGURE 8.33** Vertical displacement contours and twisting angle of numerical analyses for electrical actuation, $\mathbb{V} = 32\left(\text{kV}\right)$ and internal pressure $P_r = 1.5\left(\text{kPa}\right)$ (longitudinal deformation scale factor = 3).

However, when the fiber angle is more considerable, the axial load has a direct relation with the longitudinal stretch. As was mentioned before, in the absence of axial load, an actuator with fibers at 90 experiences a longer final length in comparison with the fiberless actuator. When some axial load is applied, the longitudinal stretch of the fiberless actuator surpasses the longitudinal stretch of the fiber-reinforced actuator.

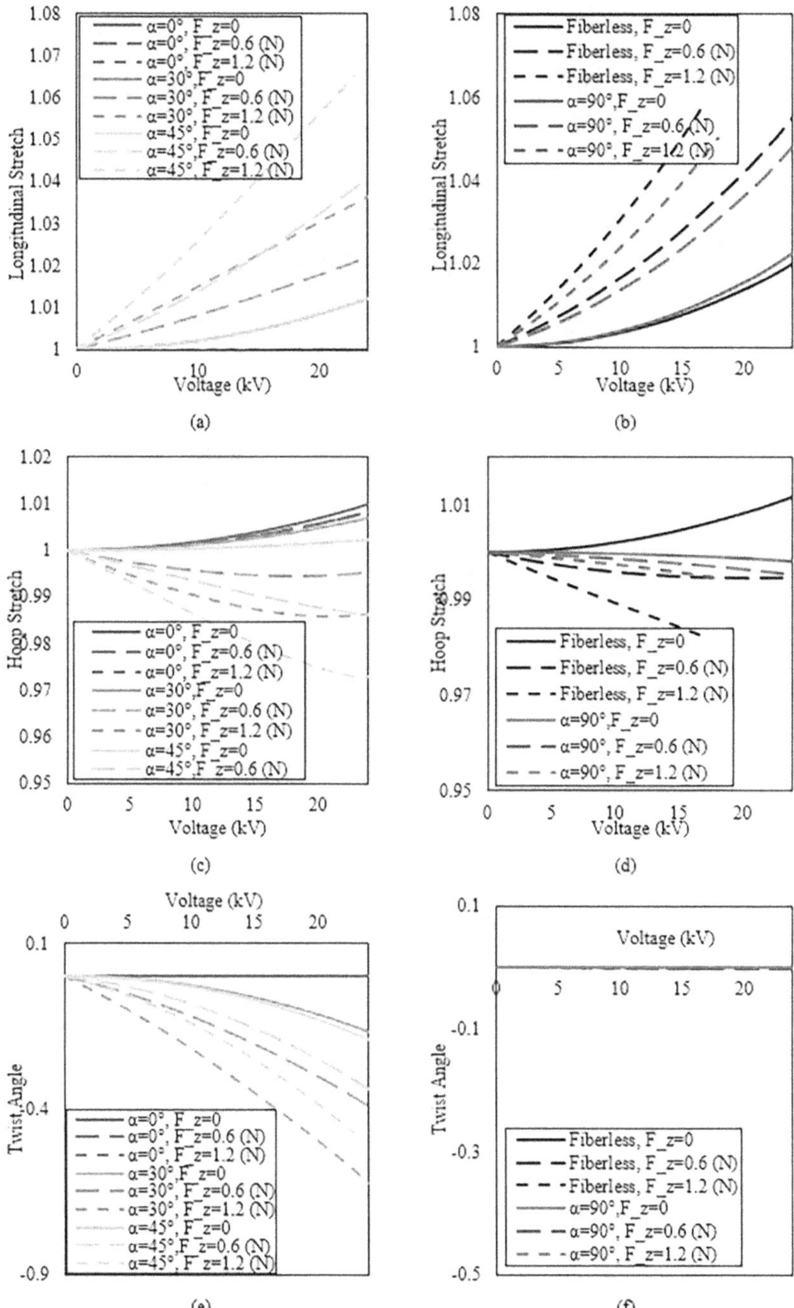

**FIGURE 8.34** Numerical result for electrical actuation and axial load (a, b) Longitudinal stretch – voltage (c, d) Hoop stretch- voltage (e, f) Twist angle – voltage.

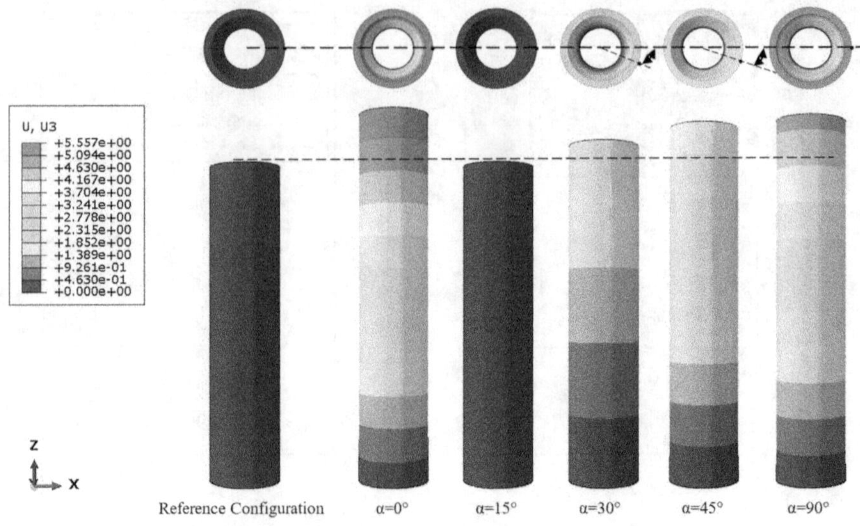


**FIGURE 8.35** Vertical displacement contours and twisting angle of numerical analyses for electrical actuation = 24 (kV) and axial load, $F_z = 0.6(\text{N})$ (longitudinal deformation scale factor = 3).

Figure 8.34(c) and (d) indicate that axial load has a negligible effect on the hoop stretch of the actuator with fibers at $\alpha = 90°$ and $\alpha = 0°$. However, for other examined actuators, the direction of the hoop stretch can change. Finally, the axial load can not cause any twist angle in axisymmetric actuators (fiberless, $\alpha = 90°$ and $\alpha = 0°$), while for actuators with fibers at $\alpha = 30°$ and $\alpha = 45°$, the effect of axial load on the twist angle is detectable. For these actuators, the higher the axial load leads, the more twisting deformation.

Vertical displacement contours and twisting angles for fiberless and fiber-reinforced actuators subjected to voltage and axial tensile force are shown in Figure 8.35.

#### 8.4.2.4 Applying Torque on TDEA

Actuators driven by electric voltage were also subjected to two different levels of torque, $T_\theta = 8(\text{N.mm})$ and $T_\theta = 16(\text{N.mm})$. The results showing the effect of torque are compared with results of mechanically unloaded actuators in Figure 8.36. Referring to Figure 8.36(a) torque application changes the direction of longitudinal deformation and cause the reinforced actuator to shorten. It has the least, and the most impacts on the longitudinal deformation of actuators stiffened with fibers at $\alpha = 0°$ and $\alpha = 30°$, respectively. On the other hand, applied torques were large enough to change the twisting direction, and as they grew, the twist angle leveled up.

Figure 8.37 displays the vertical displacement contours and twisting angles for fiberless and fiber-reinforced actuators subjected to voltage and torque.

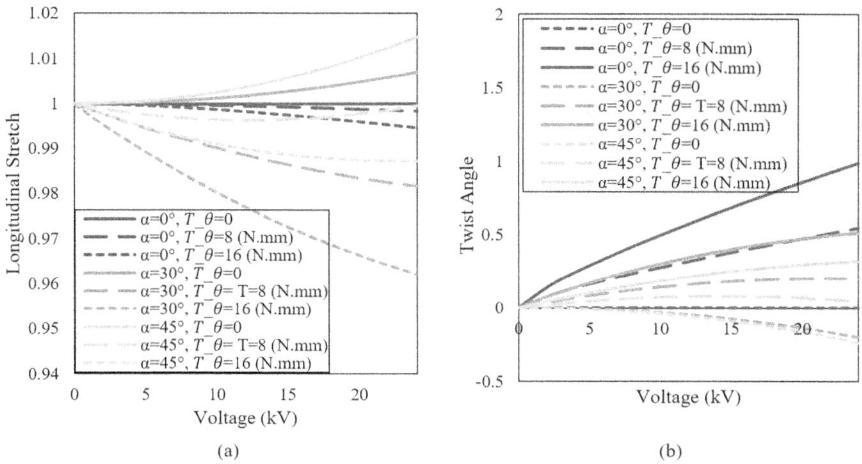


**FIGURE 8.36** Numerical result for electrical actuation and torque (a) Longitudinal stretch – voltage (b) Twist angle – voltage.

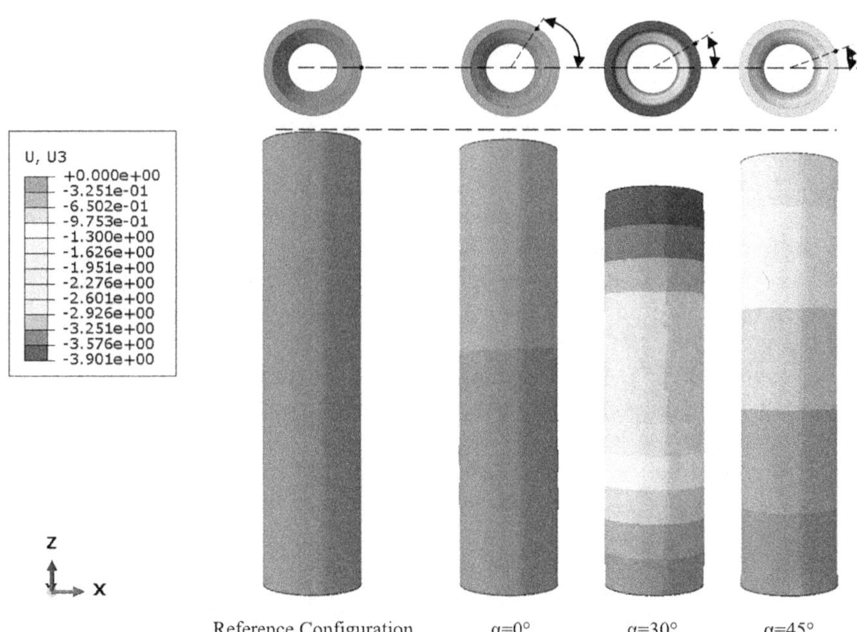


**FIGURE 8.37** Vertical displacement contours and twisting angle of numerical analyses for electrical actuation, $\mathbb{V} = 24\,(\text{kV})$ and torque, $T_\theta = 16\,(\text{N.mm})$ (longitudinal deformation scale factor = 3).

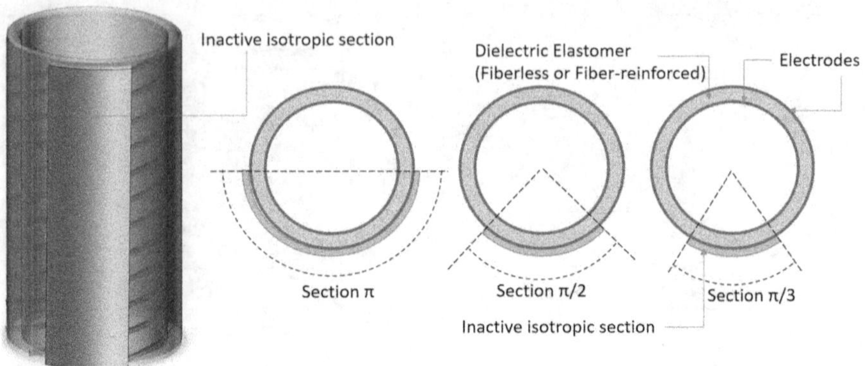


**FIGURE 8.38** Actuators with inactive isotropic attaching sections.

#### 8.4.2.5 Bending-Twisting Coupling in TDEA

Attaching an inactive isotropic hyperelastic layer (thickness = 1 mm) to the TDEA constrains the longitudinal deformation of the tube. Consequently, when an electric field is applied to the actuator, it experiences bending and twisting. To investigate its bending behavior, we simulated and attached three different sections (section $\pi$, section $\pi/2$, and section $\pi/3$) to the actuators. Further details are presented in Figure 8.38.

The presented model was utilized to simulate and analyze TDEAs with supporting sections. Numerical simulations were conducted for both fiberless and fiber-reinforced TDEAs, and the results of the actuators' bending deflection are shown in Figure 8.39.

According to the deflection results, as the inactive section enlarges and covers a bigger area of the actuator, the actuator bends more effectively, except for actuators with fibers at 0°. It is also evident that as the fibers' angles grow, application of the electric field leads to more considerable deflections. Results confirm that the bending behavior of a fiberless actuator is similar to the actuator reinforced by fibers at 90°.

Figure 8.40 shows the bending deformation and twisting angle of the simulated bending TDEAs. In addition to bending deformation, the provided figures can also confirm twisting in TDEAs. They indicate that in actuators with symmetric configurations (fiberless actuators and actuators reinforced with fibers at 0° and 90°), no twisting is detectable. However, as the fibers' angle increases from 0° to 45°, the actuators twist more.

### 8.4.3 Anisotropic Multi-Layered Dielectric Elastomer Actuator

Developed subroutine makes it possible to analyze the electromechanical behavior of multi-layered dielectric elastomers in different configurations subjected to various loadings and boundary conditions. Since the model considers each actuator feature through specific terms and constants, it is easy to assume additional

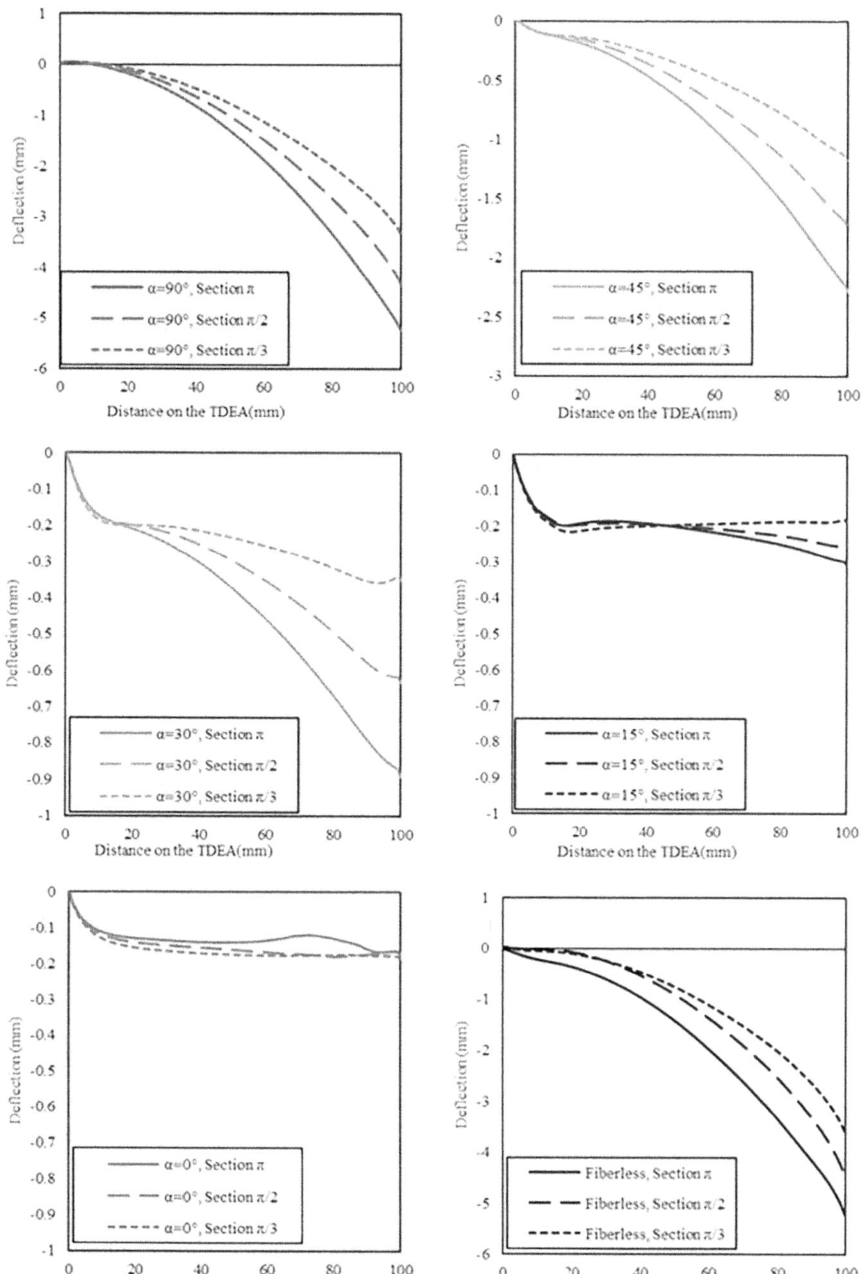


**FIGURE 8.39** Bending deflection of TDEAs with inactive isotropic attaching layers.

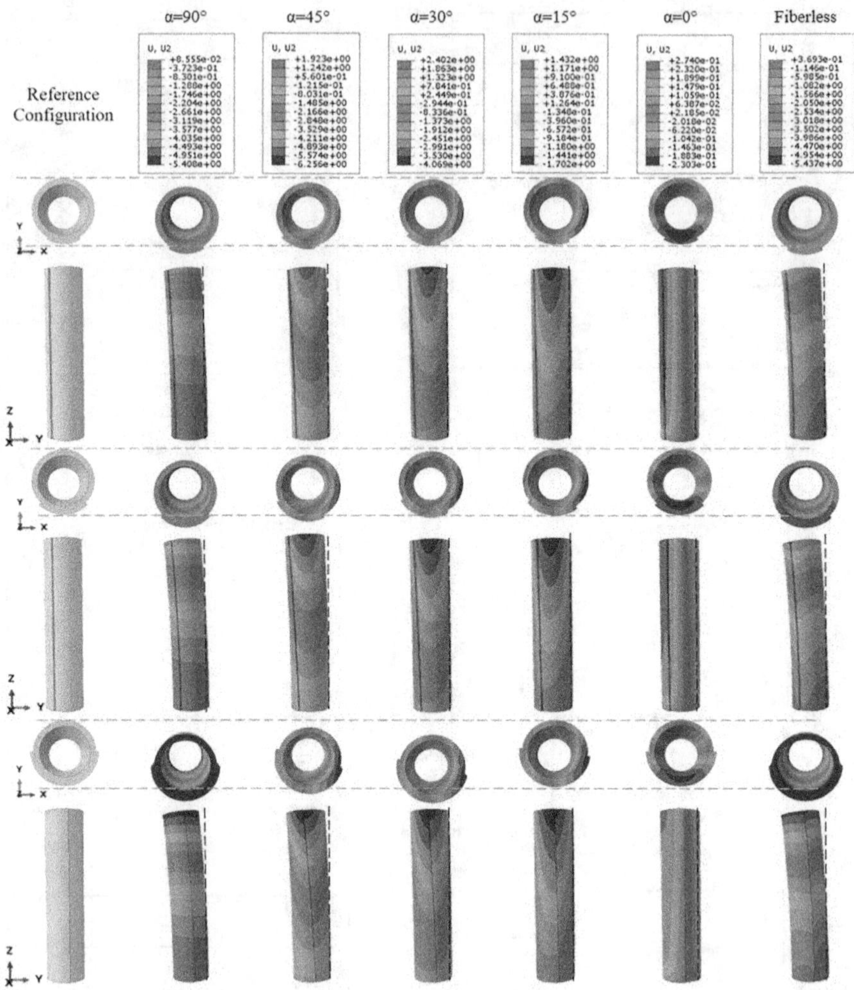


**FIGURE 8.40** Coupling of bending and twisting of TDEAs with inactive isotropic attaching layer.

attributes for the material. Moreover, there is no limitation on the number of layers in the multi-layered actuator.

Here, we cover the results of the numerical analysis of multi-layered actuators of four different configurations: (1) binder, (2) tube, (3) curved section binder, and (4) circular diaphragm. All the models are assumed to be made of VHB 4910 reinforced by one or two families of fibers.

All the geometrical parts were discretized using 8-node linear brick elements with reduced integration and enhanced hourglass control options. These options help to ensure physically realistic and accurate results by reducing locking and preventing hourglassing. Loading and boundary conditions of models are described in each section. The results are named with particular patterns to explain the layering of the model. The result names start with B (for binder), *T* (for tube), CS (for curved section binder), and CD (for circular diaphragm). The names then continue with xL, which shows the number of layers in the model and the orientation of the reinforcing fibers.

The blocking force and the deflection of the actuator as functions of actuation voltage or actuator geometry are reported for the conventional binder, curved-section binder, and circular diaphragm in the following sections. Blocking force represents the maximum force the actuator generates when it cannot move or deform. In addition to actuation displacement, blocking force is an essential parameter in the design and optimization of actuators because it is a scale to measure the performance and efficiency of the actuator.

#### 8.4.3.1 Fiber-Reinforced Multi-Layered Dielectric Elastomer Binder

The simulated binder actuator has dimensions of 100 × 10 × 1 mm. One end of the binder is entirely fixed, while the other is free to move as the electric field is applied. Three types of binders were studied: (1) single layer, (2) two layers, and (3) four layers. The whole thickness for all the models is 1 mm, and the layers of the multi-layered binder have equal thickness. Since the layers are electrically parallel, the applied voltages to the layers of a multi-layered binder are equal ($\mathbb{E} = \mathbb{V} / d$). Table 8.4 presents information for layers.

Figure 8.41 explains how the result names show the layer arrangement.

The deflection along the binder and blocking force at its tip for the multi-layered binders are presented in Figures 8.42 and 8.43 The chart depicts the deflection of the binder along its length and the blocking force-voltage relationship.

Figures 8.42 and 8.44 suggest that bilayer binders bend more when subjected to a constant voltage. However, increasing the number of layers to four in a constant thickness may not have a similar effect on the blocking force. On the other hand, in comparison with 2-layer binders, 4-layer binders generally produce a more significant blocking force, except for B-2L, 30-45 and B-4L,30-45-30-45.

Moreover, according to the results, the bending deflection and blocking force of B-2L, *x-y*, and B-2L, *y-x* are opposite.

According to the results, 2-layer and 4-layer binders experience about 35 to 45 times the bending actuation compared with single-layer binders (depending on the fiber orientations). This finding implies that multi-layered binders need lower

**TABLE 8.4**
**Thickness and Electric Field of Each Layer in Different Multi-Layered Binder**

| | Single Layer | 2 Layers | 4 Layers |
|---|---|---|---|
| Thickness of layer, *d* (mm) | 1 | 0.5 | 0.25 |
| Electric field, $\mathbb{E}$, (kV/mm) | 4 | 8 | 16 |

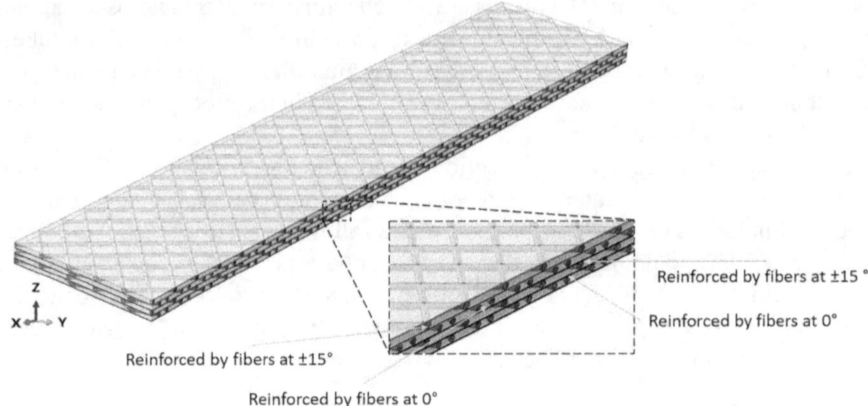


**FIGURE 8.41** Dielectric elastomer fiber-reinforced multi-layered binder, named B-4L, 0-15-0-15.

actuation voltage to induce the same displacement and force than single-layer ones. Therefore, multi-layered structures provide a safer and more stable application.

Figure 8.45 depicts the displacement contour for one of the four-layer binders and the corresponding bilayer and single-layer actuators.

#### 8.4.3.2 Multi-Layered Anisotropic Tubular Dielectric Elastomer Actuator

The next studied model is a tubular actuator with an outer radius, length, and thickness of 10.5, 100, and 2 mm, respectively. One end of the tube is fixed, while the other can deform. This actuator is assumed to be made of one or two layers in different analyses. Two layers of a bilayer tubular actuator have a similar thickness of 1 mm. An electric field along the thickness of the tube actuates the model, and the actuation voltage of the model is 4 kV. It is worth noting that the layers of the tubular actuator are assumed to be reinforced by one family of fibers. Figure 8.46 depicts the code name of the analysis according to the layers' arrangement.

According to Figure 8.47, the longitudinal stretches of all the tubular actuators are less than 1.01. T-1L,90, T-2L,45-90, and T-2L,30-90 undergo the highest longitudinal stretches. This implies that adding a layer with fibers at an angle of less than 90° constrains the longitudinal stretch. The results also show that using fibers at different angles cannot decrease the effect of fibers at 0° in limiting longitudinal stretch. Any other combination of stiffening fiber angles causes a longitudinal stretch between the longitudinal stretches of the single layer with that fiber angle. For example, the longitudinal stretch of T-2L,30-45 is less and higher than that of T-1L,45, and T-1L-30, respectively.

Figures 8.48 and 8.49 display the effect of layering up and fiber orientation on the hoop stretch and twist angle of the tubular actuator. Figures 8.48(a) and 8.49(a) depict that as long as the fiber angles in the layers of a bilayer tubular actuator are equal, the layer arrangement would not create a considerable difference in hoop

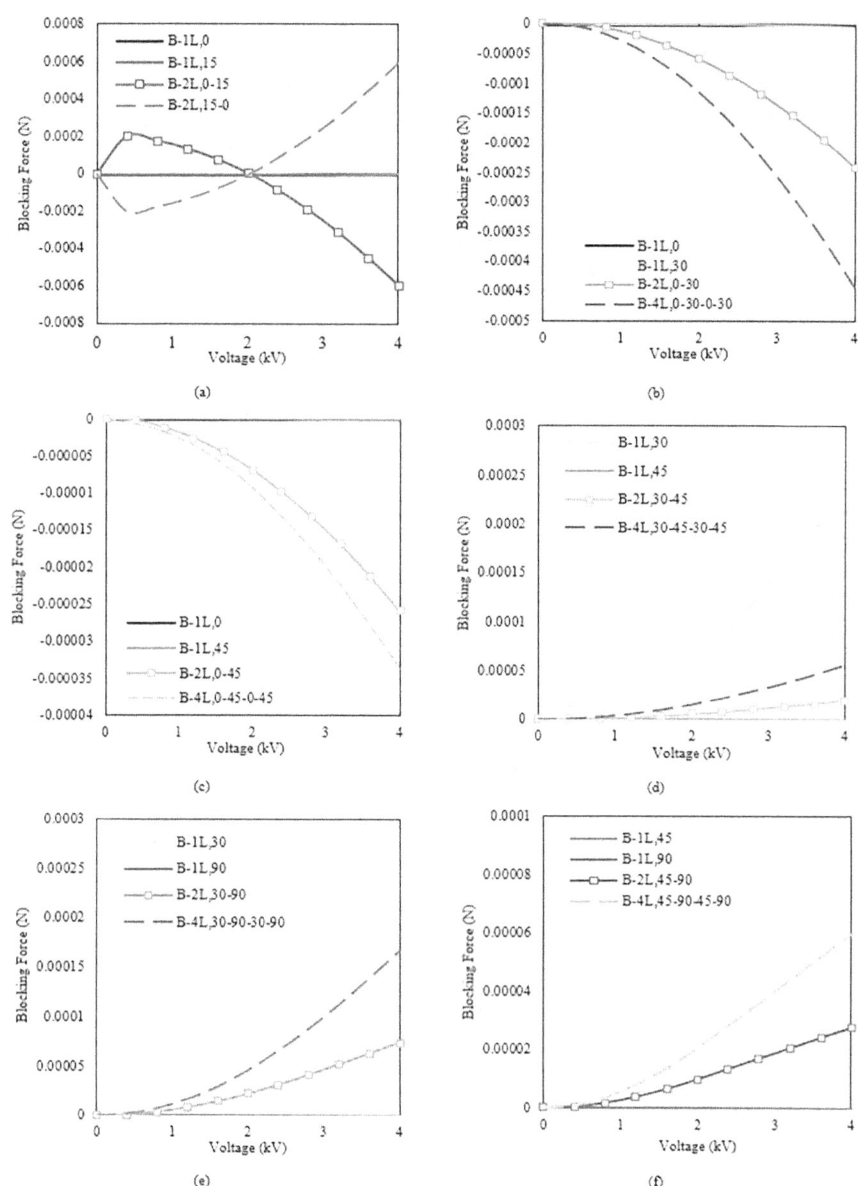


**FIGURE 8.42** Blocking force of dielectric elastomer fiber-reinforced multi-layered binder.

stretch and twist angle (same as longitudinal stretch). Figure 8.49 also conveys that the hoop stretches of a bilayer made of layers with different fiber angles is between the hoop stretch of the corresponding single layer. The twist angle of the layer reinforced by fibers at 0° is slight. However, combining the layer with fibers

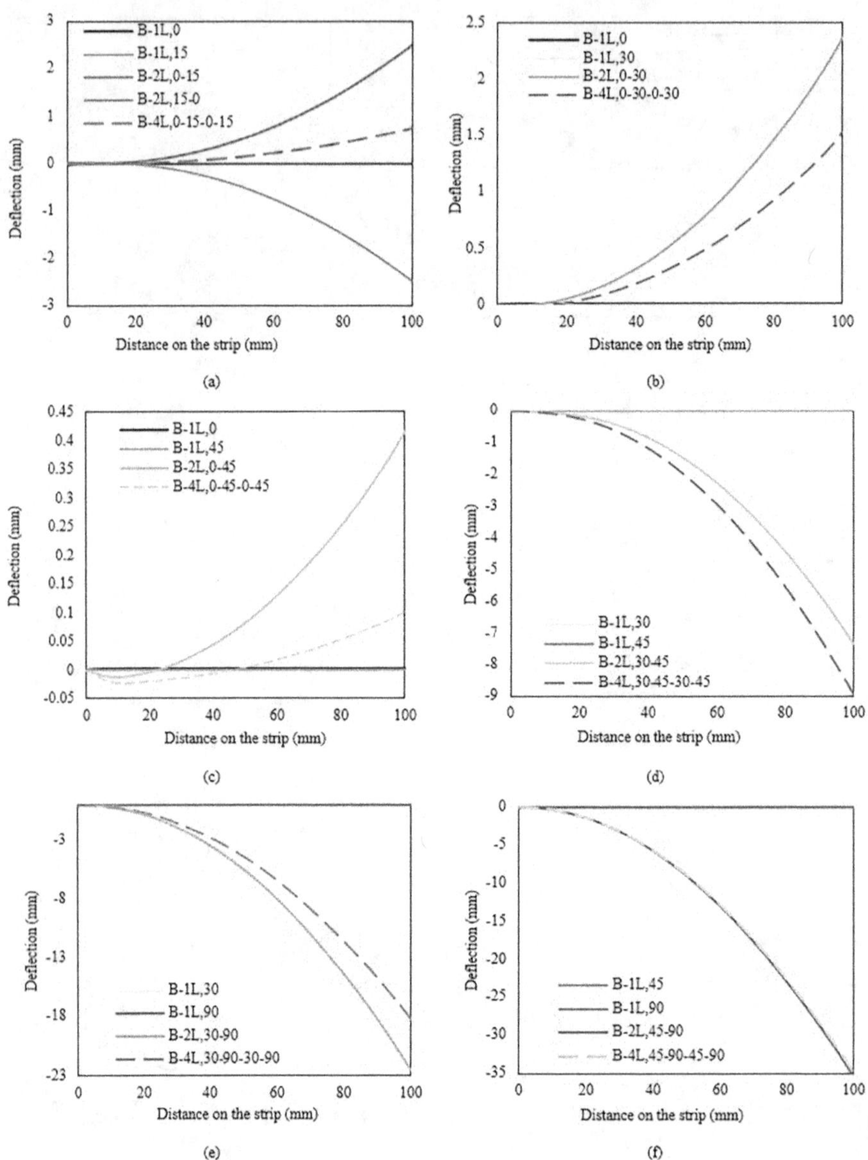


**FIGURE 8.43** Deflection of dielectric elastomer fiber-reinforced multi-layered binder.

at another orientation can induce twisting in the tubular actuator. It is also worth noting that during the actuation of some bilayer arrangements, the actuator may twist in opposite directions, which is not the case for single layers.

These results suggest that layering with fibers at different orientations may change the longitudinal-twisting coupling degree. For example, if a single layer

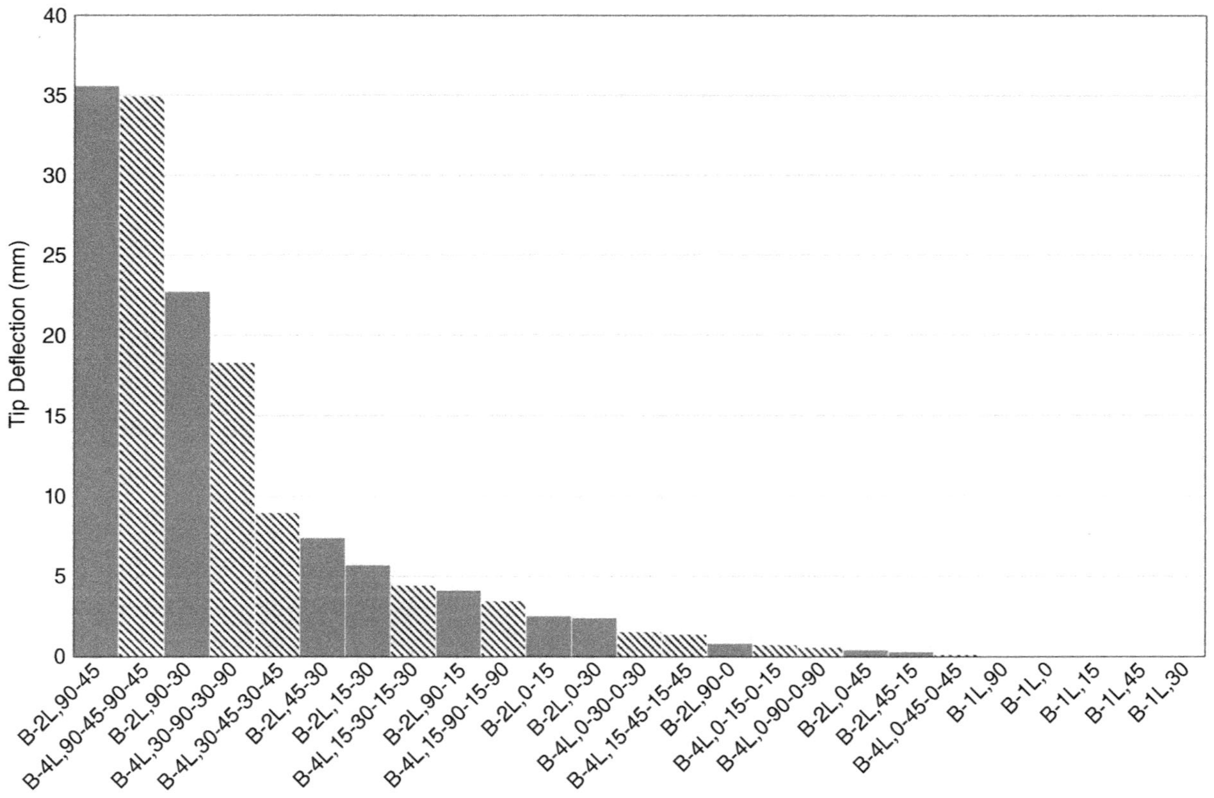


**FIGURE 8.44** Tip deflection of dielectric elastomer fiber-reinforced multi-layered binder.

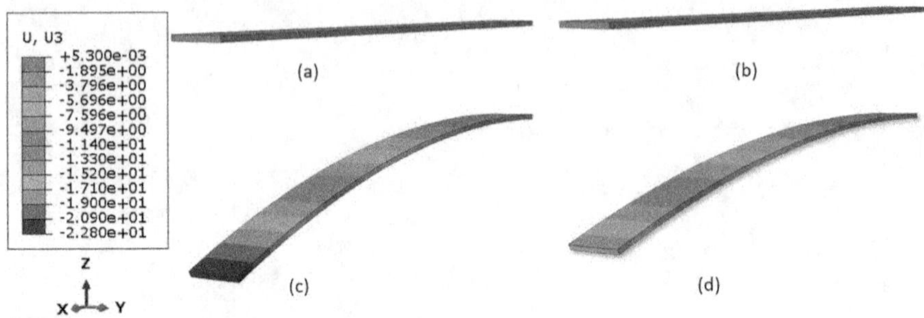


**FIGURE 8.45** Deflection contour for dielectric elastomer fiber-reinforced multi-layered binder (a) B-1L,30 (b) B-1L,90 (c) B-2L,30,90 (d) B-4L, 30,90,30,90.

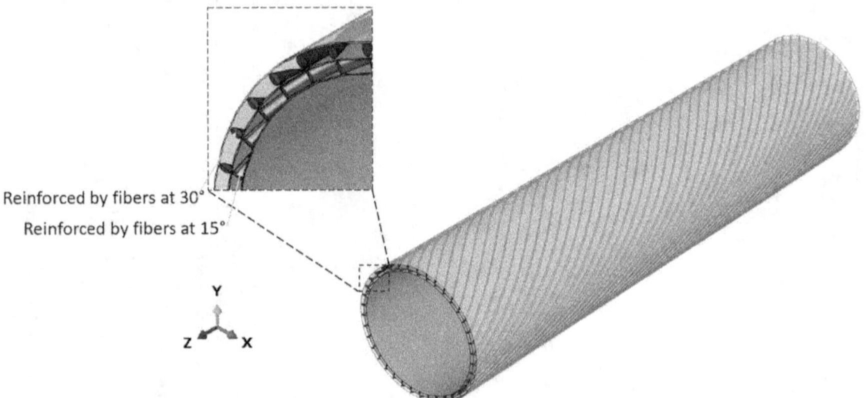


**FIGURE 8.46** Multi-layered anisotropic tubular dielectric elastomer actuator, named T-2L, 30-15.

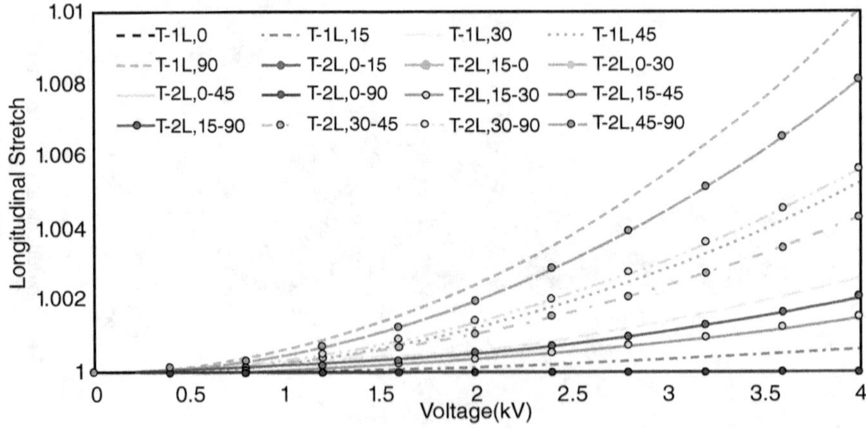


**FIGURE 8.47** Longitudinal stretch of multi-layered anisotropic tubular dielectric elastomer actuator.

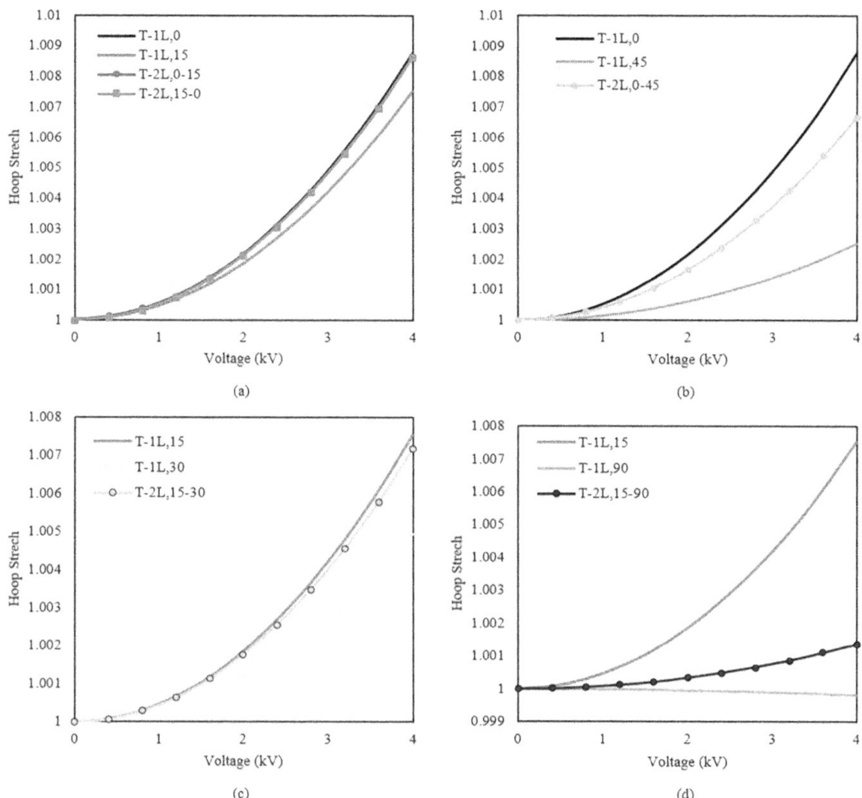


**FIGURE 8.48** Hoop stretch twist angle of multi-layered anisotropic tubular dielectric elastomer actuator.

with fibers at 15° is replaced with a single layer with fibers at 90°, its longitudinal stretch and hoop stretch increase and decrease, respectively, while its twist angle declines to about zero. However, replacing that single layer with a bilayer tubular actuator with embedded fibers at 15° and 90° not only increases the longitudinal stretch but also limits the hoop stretch decrease and causes a twisting angle with the same value in the opposite direction during the actuation. Such a mechanism provides a valuable tool for designing unique actuators based on specific needs in various fields. Figure 8.50 displays how layering up can affect the horizontal deformation of a tubular actuator.

#### 8.4.3.3 Multi-Layered Anisotropic Dielectric Elastomer Curved-Section Binder

We designed another model to combine the unique properties of a conventional bending actuator and a tubular actuator. The length, outer radius, inner radius, and thickness of the curved section binder are 100, 9.5, and 1 mm, respectively. The 2-layer binder consists of two layers of equal thickness. A 4 kV voltage actuates

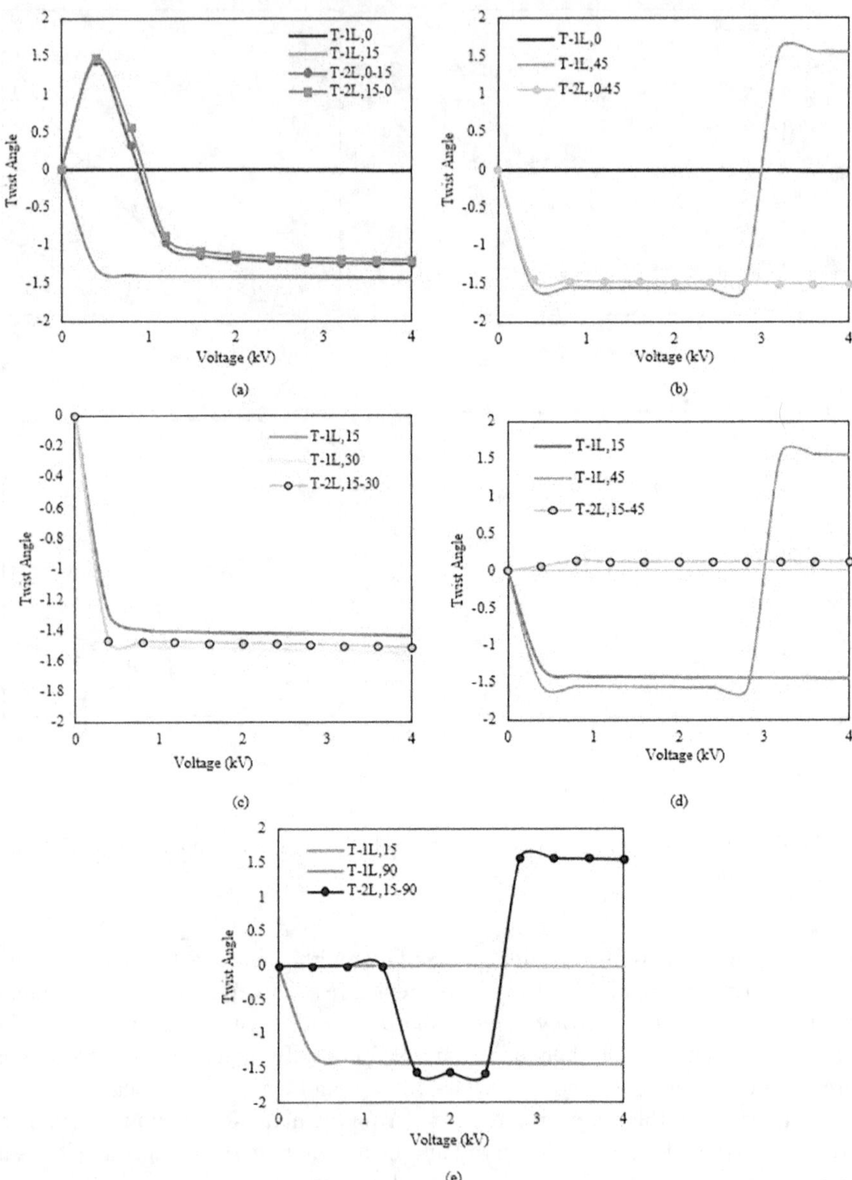


**FIGURE 8.49** Twist angle of multi-layered anisotropic tubular dielectric elastomer actuator.

each actuator. To understand the naming pattern and layer arrangement, please see Figure 8.51. The layers of this actuator are also reinforced by one family of fibers.

As seen in Figure 8.52, the curved section binder with fibers at 0° (longitudinal fibers) has the slightest bending deflection along the strip. As the fiber angle increases, the deflection of the binder, actuated by the same voltage, increases.

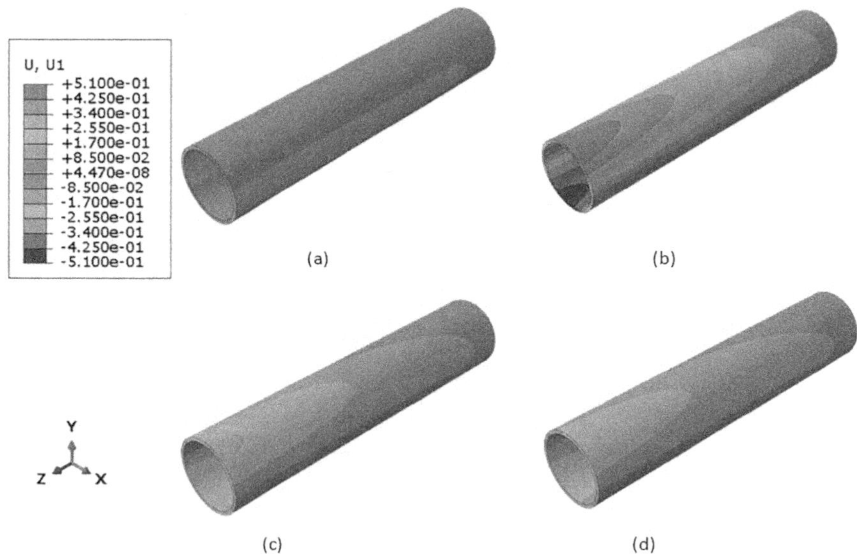


**FIGURE 8.50** Horizontal displacement contour for multi-layered anisotropic tubular dielectric elastomer actuator (a) T-1L,0 (b) T-1L,15 (c) T-2L,0,15 (d) T-2L,15,0.

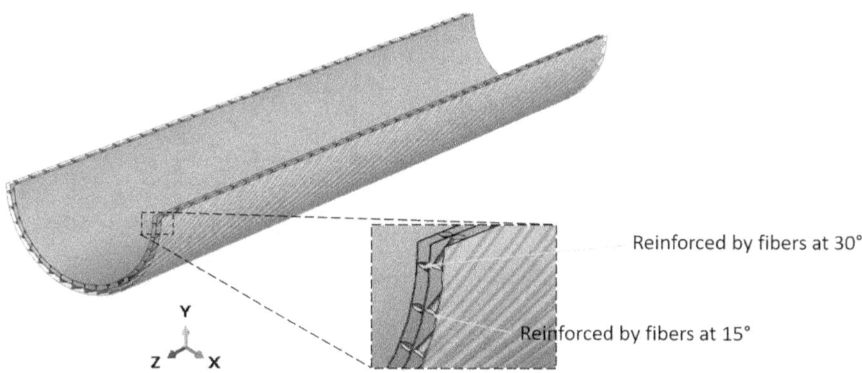


**FIGURE 8.51** Multi-layered anisotropic dielectric elastomer curved section binder, named CS-2L,15-0.

According to the results shown in Figures 8.53 and 8.54, adding layers with fiber in different directions cannot change the bending behavior of a single-layer curved-section binder with fibers at 0°. However, it completely changes the blocking force (measured at the tip of the binder) of the binder.

Figures 8.53 and 8.54 show that while the layer arrangement does not affect the behaviors of a bilayer made of layers with fibers at 0° and 45°, it is a leading factor for blocking force. When the bilayer has fibers at other layers, in addition to

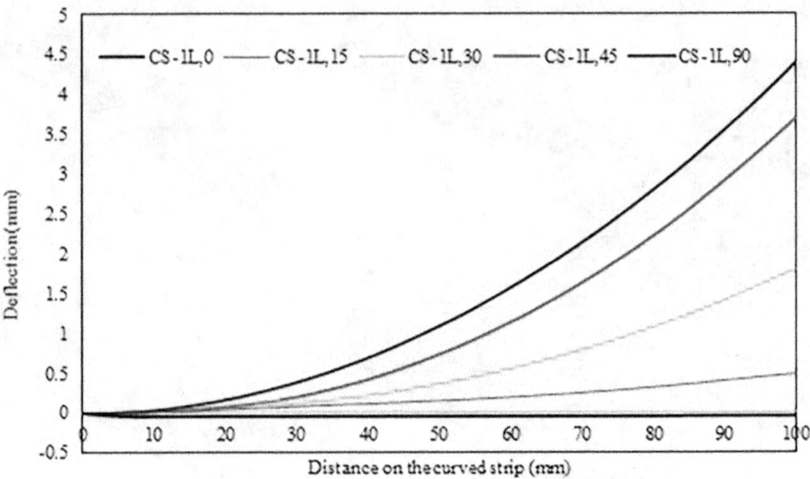


**FIGURE 8.52** Deflection of single-layer anisotropic dielectric elastomer curved section binder.

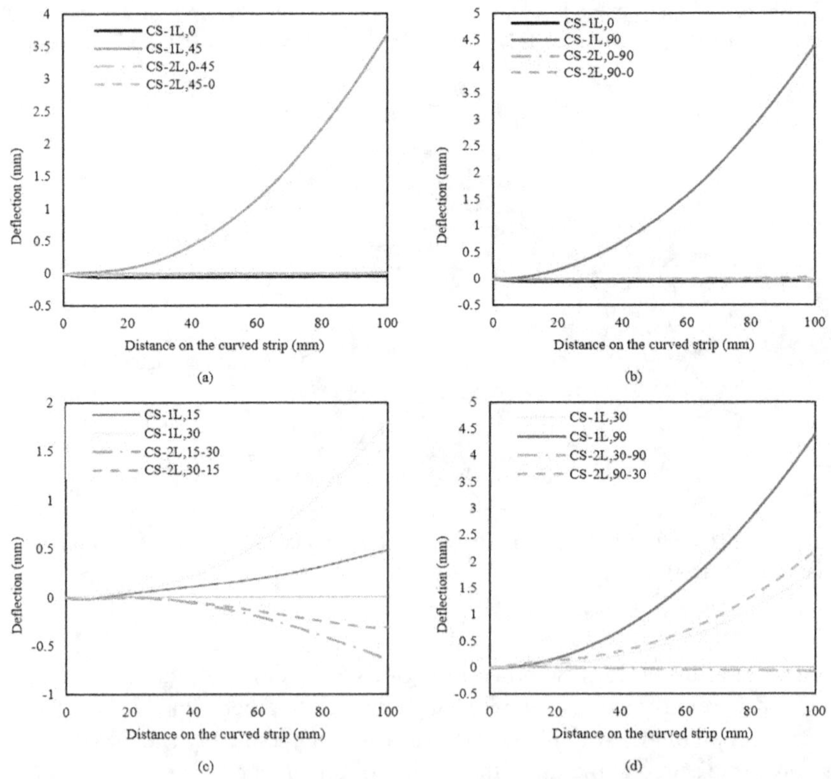


**FIGURE 8.53** Deflection of the multi-layered anisotropic dielectric elastomer curved section binder.

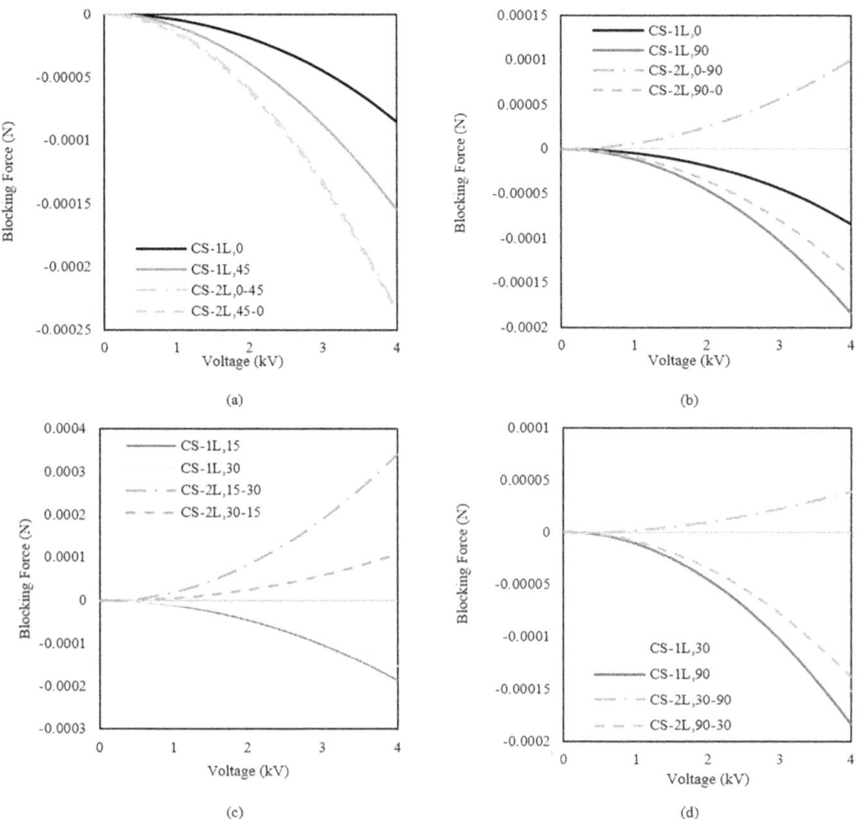


**FIGURE 8.54** Blocking force of the multi-layered anisotropic dielectric elastomer curved section binder.

deflection values, layering up may even change the deflection orientation of the curved-section binder.

Depending on the fiber direction of the added layer, blocking force and deflection may or may not vary by the layer arrangement. Similar to a conventional binder, the binder with the highest amount of deflection does not have the most significant blocking force. It is also seen that the bending of bilayers, compared with the single-layer actuator, does not follow a specific rule.

According to the chart in Figure 8.55, most bilayers have less maximum deflection than both or at least one of the single layers with corresponding fiber orientations. This fact highlights the importance of layers arrangement in bending of curved-section binder.

A curved-section binder experiences less change in bending behavior by layering up compared with a conventional binder. The deflection of a curved section binder in four different configurations and layer arrangements is displayed in Figure 8.56.

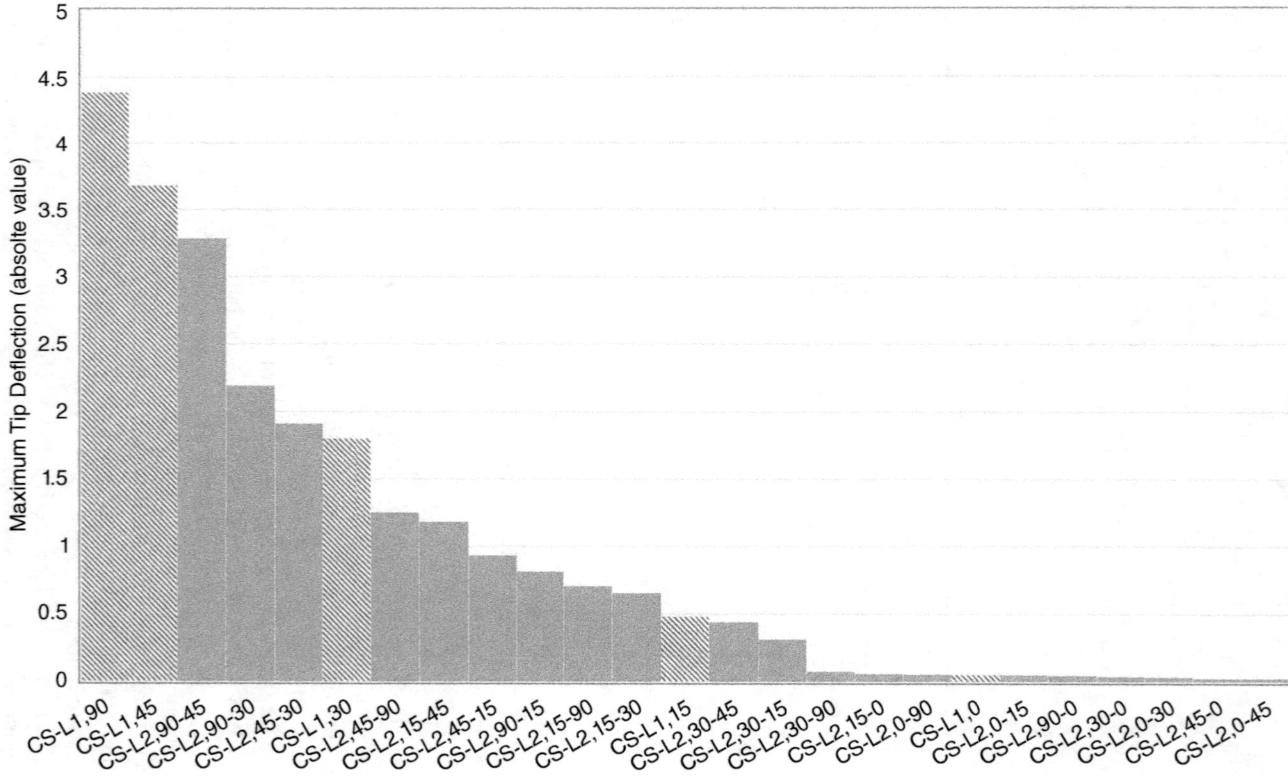


**FIGURE 8.55** Maximum deflection of multi-layered anisotropic dielectric elastomer curved section binder.

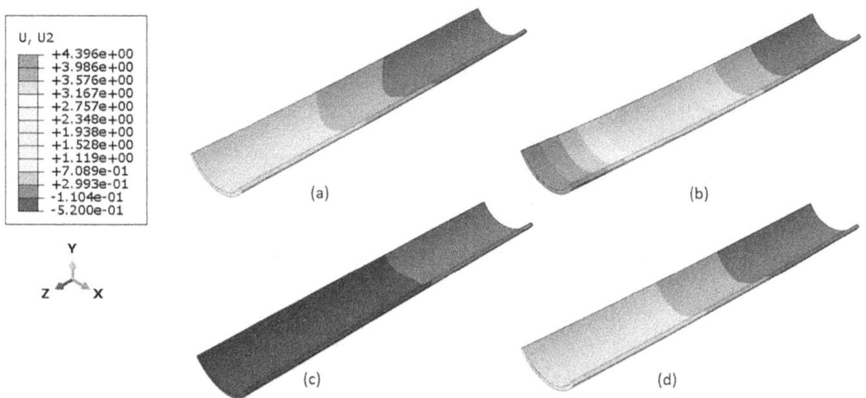


**FIGURE 8.56** Deflection contour for multi-layered anisotropic dielectric elastomer curved section binder (a) CS-1L,30 (b) CS-1L,90 (c) CS-2L,30,90 (d) CS-2L, 90,30.

#### 8.4.3.4 Multi-Layered Anisotropic Dielectric Elastomer Circular Diaphragm

A multi-layered circular diaphragm consisting of one, two, and four layers is studied in different analyses, too. The radius and thickness of each layer are 100 and 0.05 mm, respectively. For diaphragms with more than one layer, the thickness is equally divided among the layers. The outer circumference of the diaphragm is entirely fixed, and each layer is actuated by a 5.5 (kV/mm) electric field. The diaphragm analysis names follow the pattern explained in Figure 8.57.

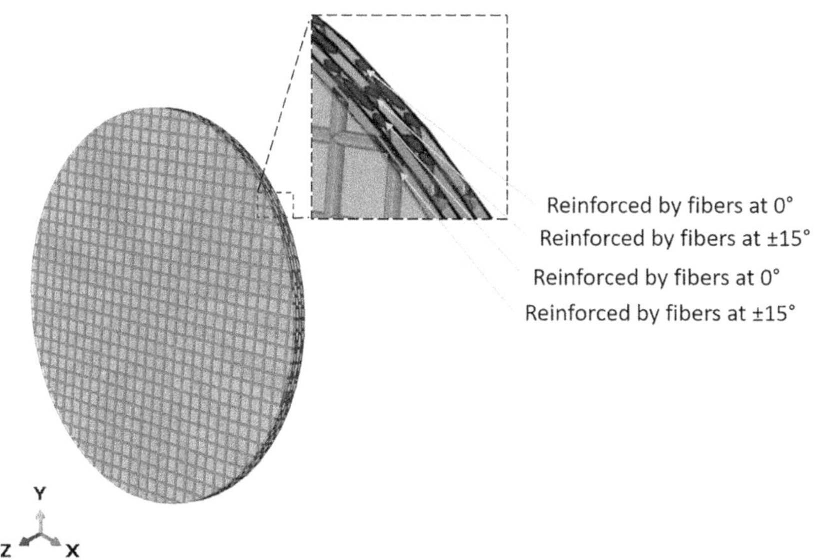


**FIGURE 8.57** Multi-layered anisotropic dielectric elastomer circular diaphragm, named CD-4L, 0-15-0-15.

The displacement and blocking force in the center of the diaphragm are plotted versus the applied voltage along the thickness in Figures 8.58 and 8.59. The blocking force of this actuator is extracted in analyses when the out-of-plane displacement at the center of the diaphragm is constrained.

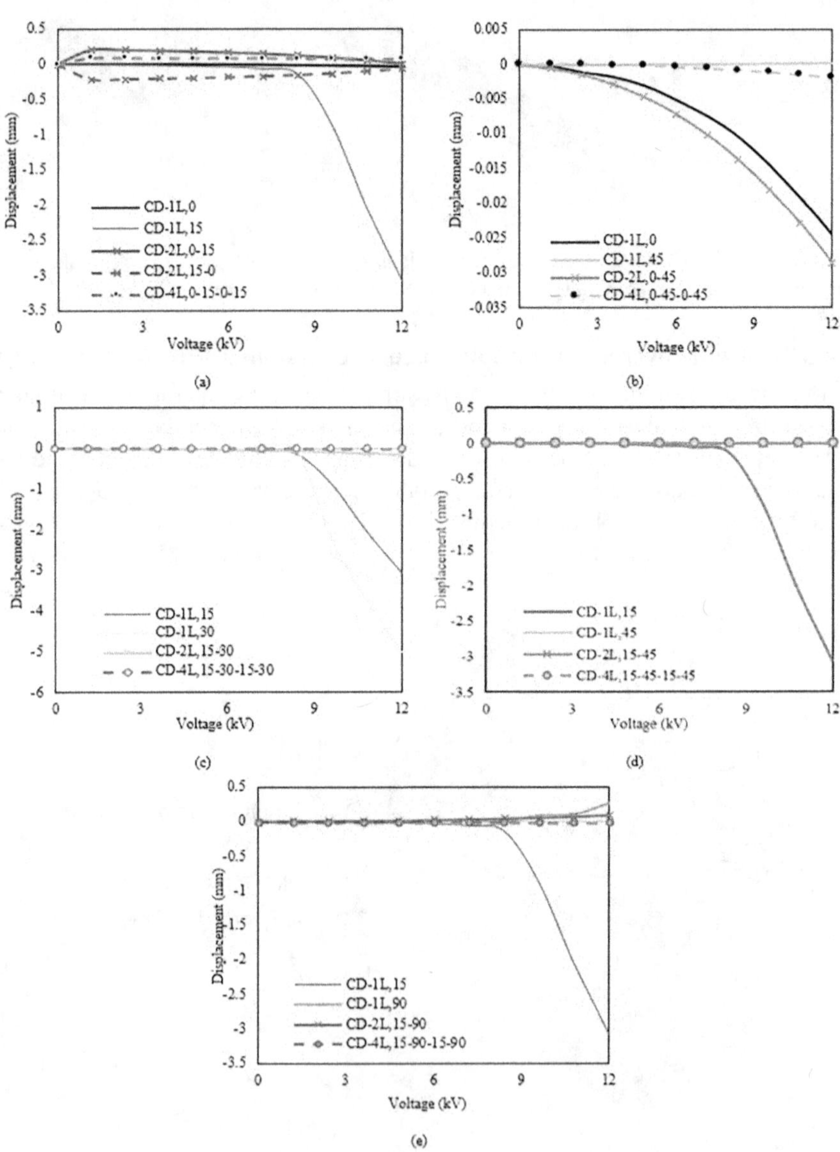


**FIGURE 8.58** Deflection at the center of multi-layered anisotropic dielectric elastomer circular diaphragm.

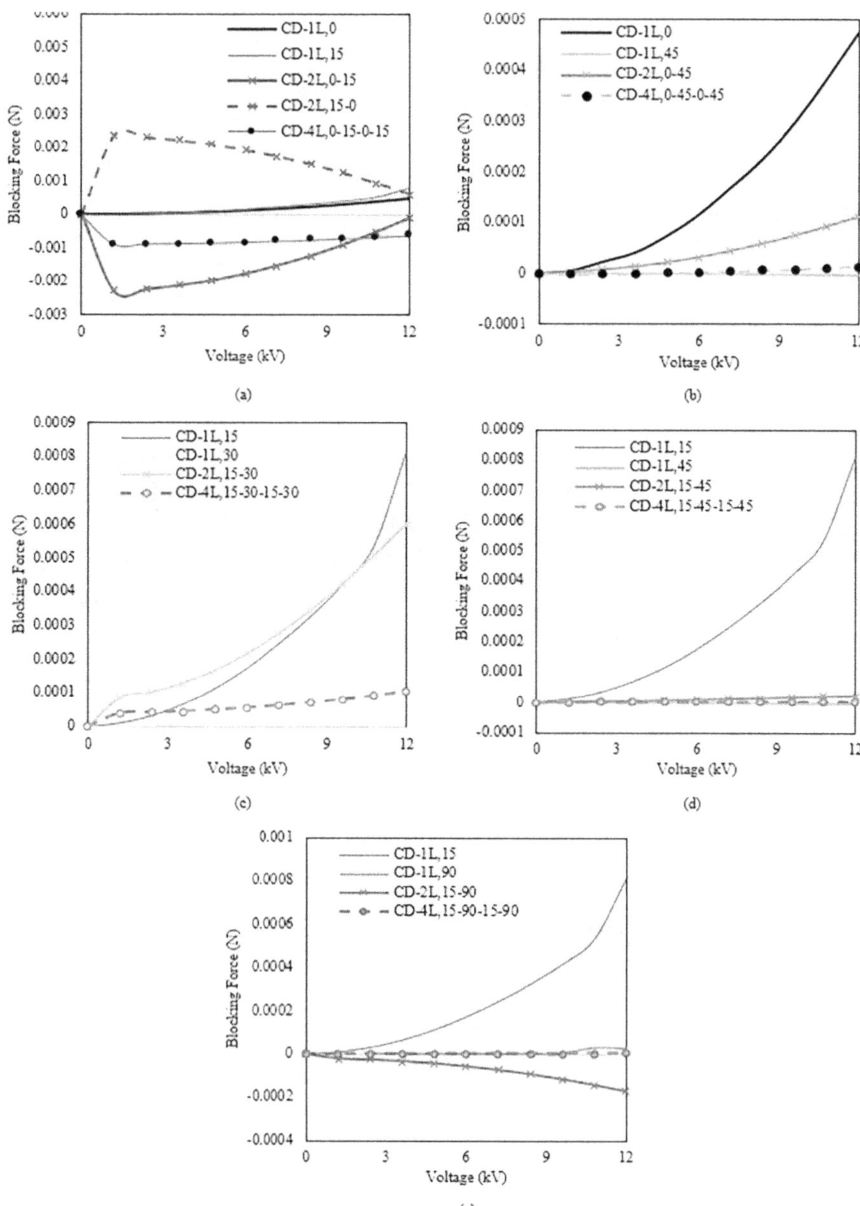


**FIGURE 8.59** Blocking force at the center of multi-layered anisotropic dielectric elastomer circular diaphragm.

Results showed that changing the order of layers with fibers at different orientations in a bilayer diaphragm leads to opposite values of force and displacement. The out-of-plane displacement of bilayer diaphragms, compared with single-layer ones, does not follow a constant rule. Depending on the orientation of the

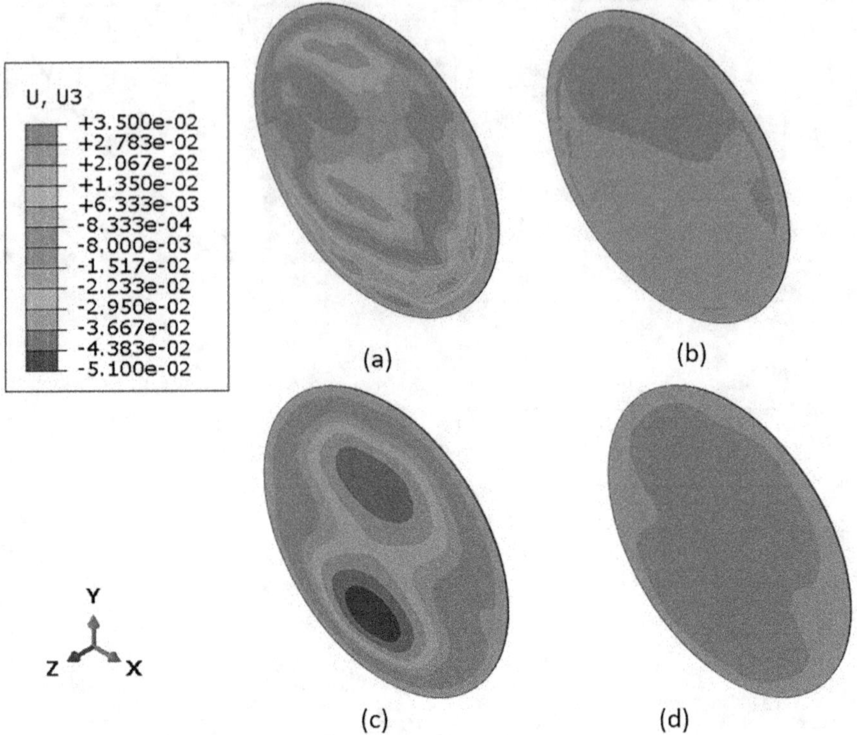


**FIGURE 8.60** Out-of-plane displacement contour for multi-layered anisotropic dielectric elastomer circular diaphragm (a) CD-1L,0 (b) CD-1L,45 (c) CD-2L,0,45, (d) CD-4L, 0,45,0,45.

embedded fibers in the layers, the bilayer diaphragm may have a bigger, smaller, or equal amount of displacement in corresponding single-layer diaphragms. However, increasing the number of consisting layers to four, in constant thickness, always lessens the amount of center out-of-plane displacement. The blocking force of two-layer and four-layer diaphragms may or may not have different behavior depending on the fibers' orientation in the layers.

The out-of-plane displacement of a circular diaphragm in four configurations, shown in Figure 8.60, conveys some effects of layering up and layer arrangement.

## 8.5 NONLINEAR DYNAMICS ANALYSIS OF ANISOTROPIC DIELECTRIC ELASTOMERS

Solving the equation of motion for a uniaxial tensile loading when the constitutive laws are derived using Neo-Hookean, Mooney-Rivlin and Ogden hyperelastic models lead to Figures 8.61–8.65. The figures present strain-time and strain-strain rate results for different fiber directions. based on the diagrams the nonlinear behavior of the structure is quite evident.

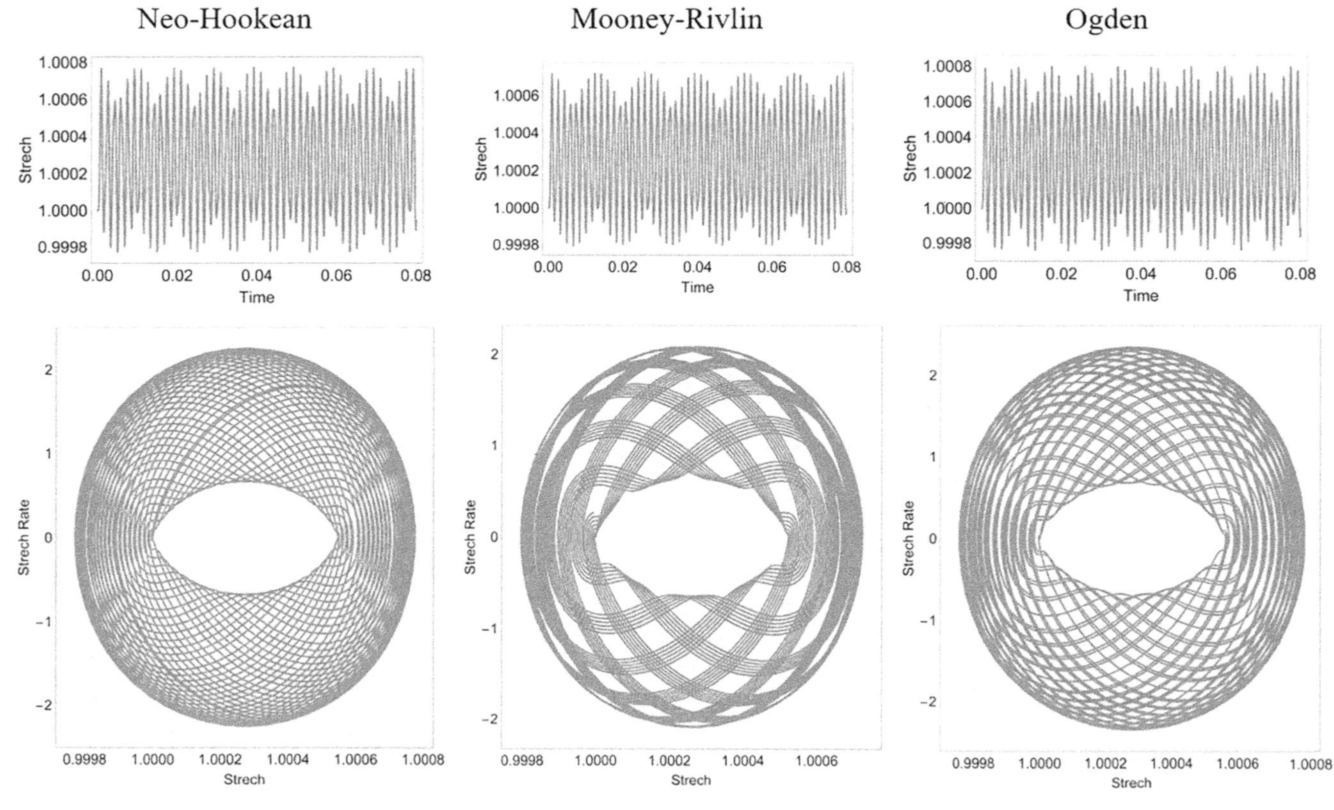


**FIGURE 8.61** Phase diagrams and oscillations of various models for uniaxial tension and $\alpha = 0°$.

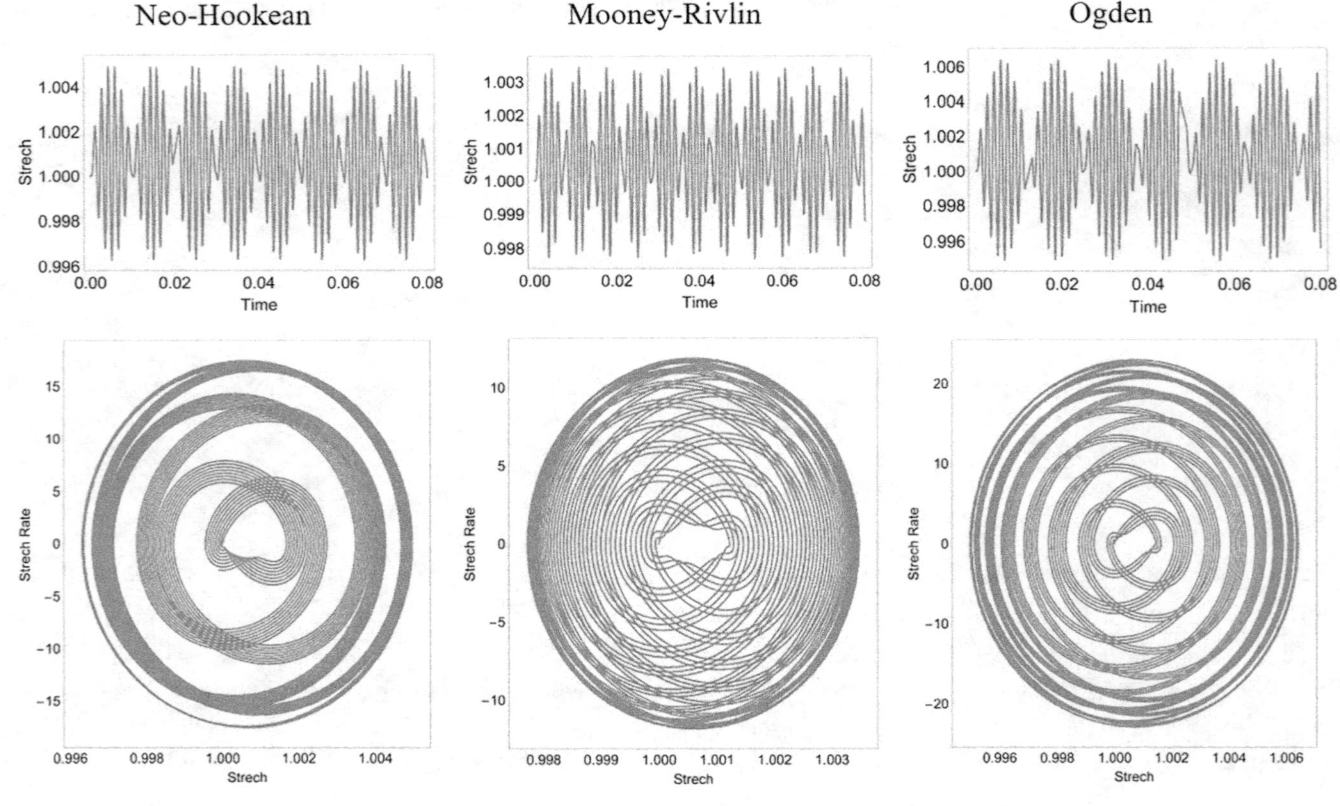


**FIGURE 8.62** Phase diagrams and oscillations of various models for uniaxial tension and $\alpha = 30°$.

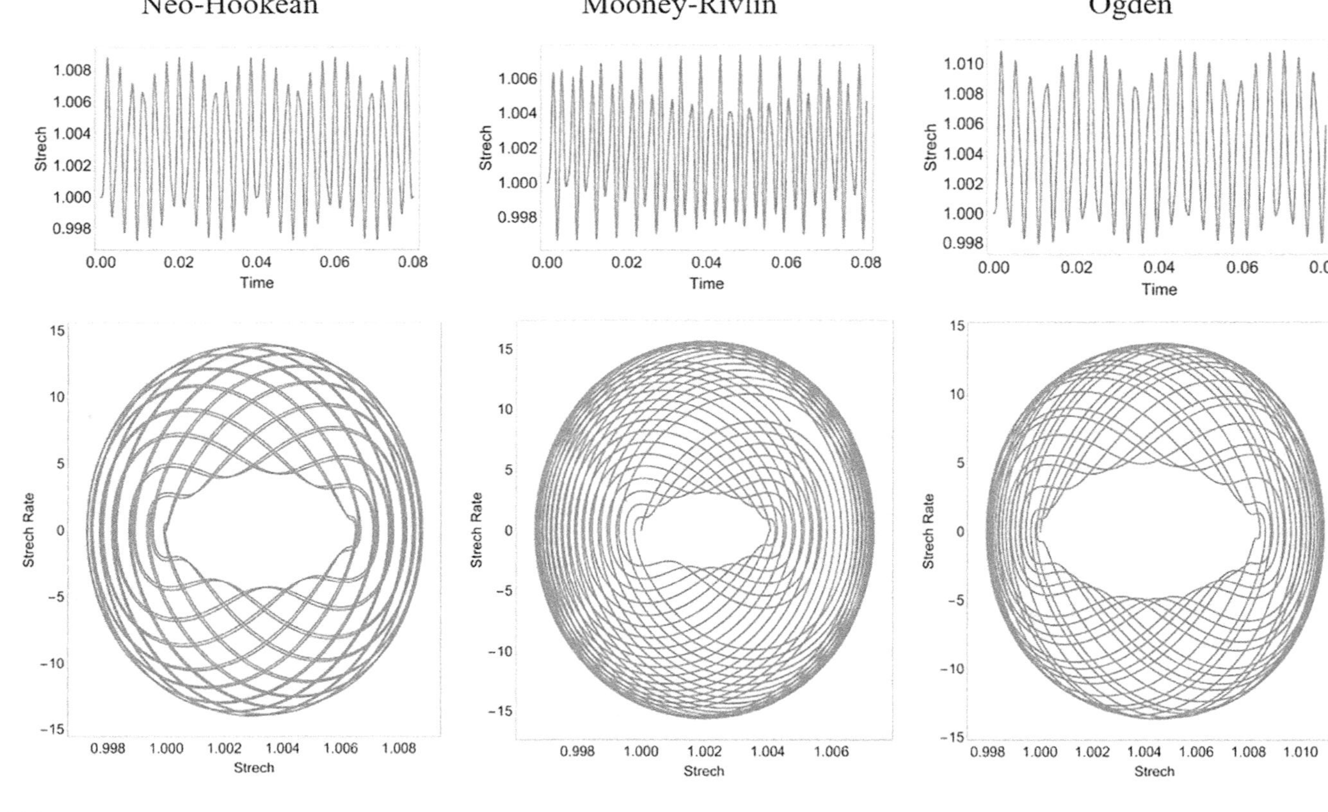


**FIGURE 8.63** Phase diagrams and oscillations of various models for uniaxial tension and $\alpha = 45°$.

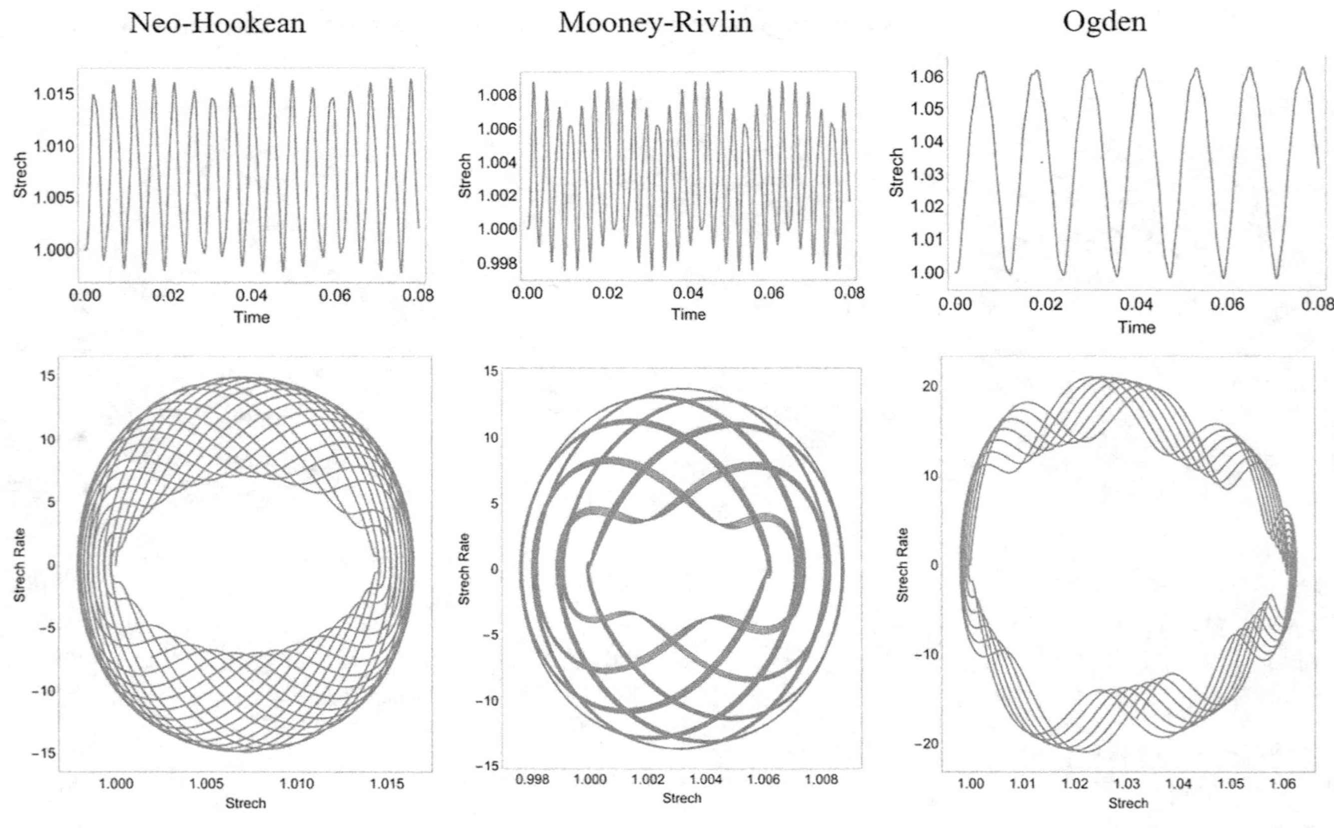


**FIGURE 8.64** Phase diagrams and oscillations of various models for uniaxial tension and $\alpha = 60°$.

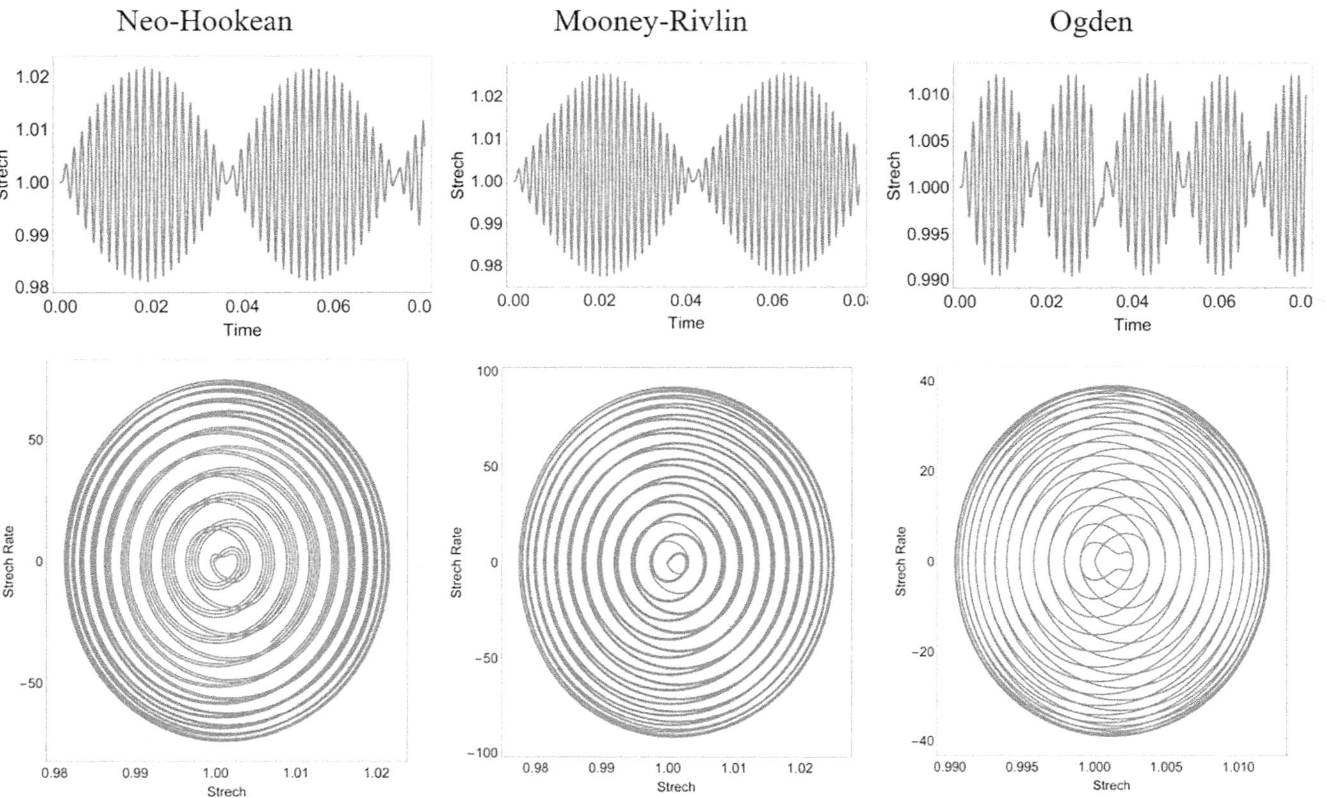


**FIGURE 8.65** Phase diagrams and oscillations of various models for uniaxial tension and $\alpha = 90^{\circ}$.

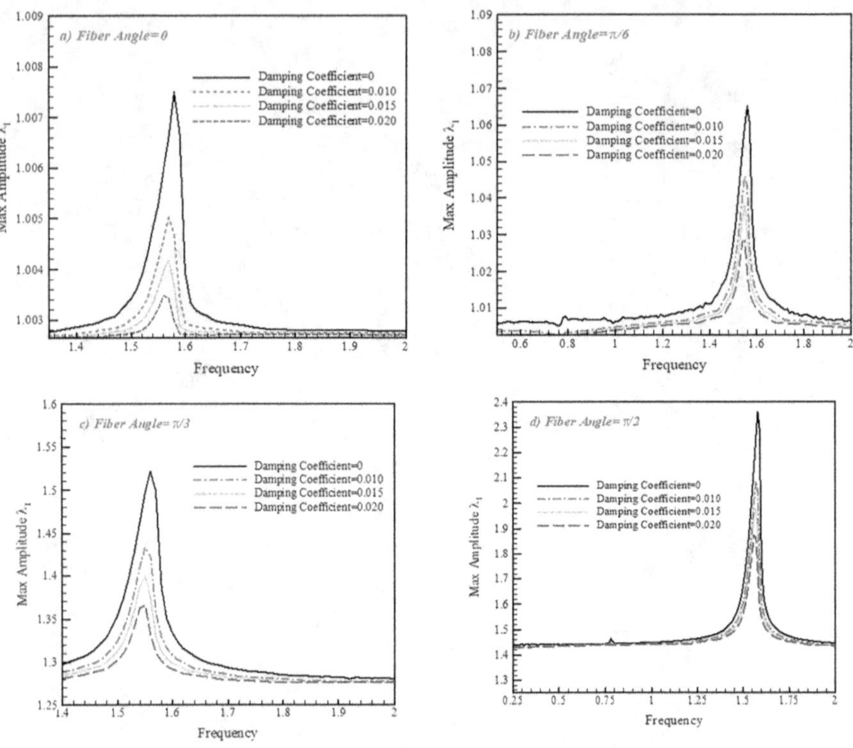


**FIGURE 8.66** Amplitude of $\lambda_1$ and frequency response curves with damping coefficients $c = 0,0.01, 0.015, 0.02$ for different fiber angle orientation (a) $\alpha = 0°$, (b) $\alpha = 30°$, (c) $\alpha = 60°$, (d) $\alpha = 90°$.

Based on Equations (5.85) and (5.86) for a time-dependent input voltage as, $\mathbb{V}(t) = \mathbb{V}_{dc} + \mathbb{V}_{ac}\sin(\Omega t)$, Figures 8.66 and 8.67 plot the amplitude (maximal values of stretch) versus induce frequency. It can be concluded that for both max amplitude the effect of damping is highly significant and by increasing this coefficient, the maximal amplitude becomes smaller. Also, the specific effect of fiber angles cannot be ignored. For example, when $\alpha = 90°$ the maximal value of $\lambda_1$ for $c = 0.015$ is almost 2, but the maximal value of $\lambda_2$ for the same fiber angle and same damping coefficient is 1.004. The reason is that because the angle of the fiber is 90 degrees, it is easier to move and change the amplitude in the axial direction, and this value is a large number. In contrast in the other direction, the fiber is entirely in the transverse direction and prevent moving so the amplitude does not change quickly. Furthermore, at the resonant frequency $\Omega = 1.57$ and $\alpha = 30°$ the maximal value of $\lambda_1$ regardless of damping coefficient is smaller than $\lambda_2$ with the same values of fiber orientation and resonant frequency. By increasing the fiber orientation, the opposite results are obtained. Which is that the maximal value $\lambda_1$ is larger than $\lambda_2$ regardless of the damping coefficient. As was mentioned before, the same behavior is seen in Figure 8.67 that gradually the maximal amplitude $\lambda_2$ reduces with a constant resonant frequency and for all different fibers directions.

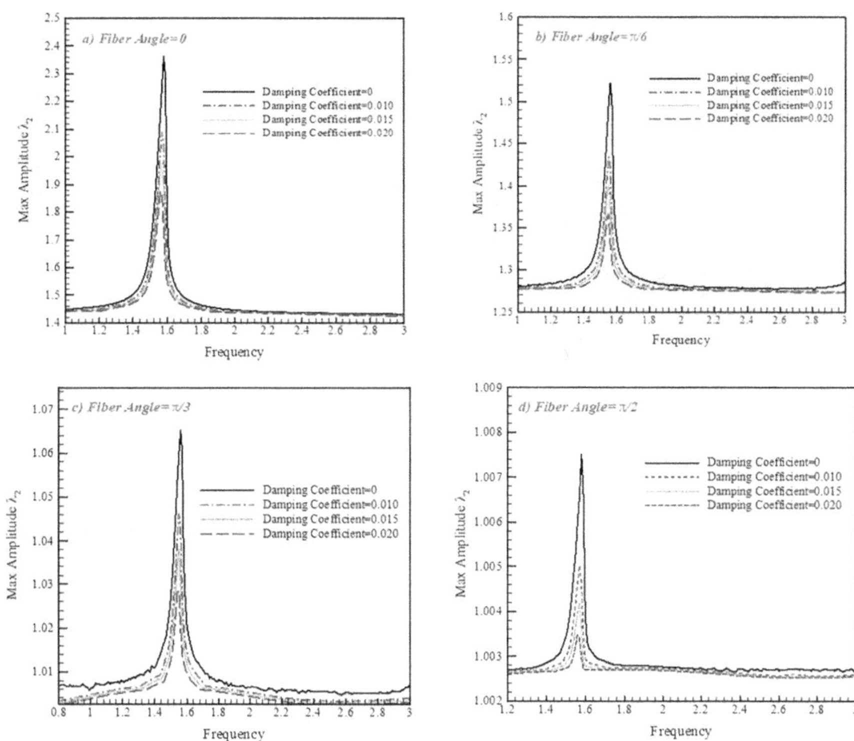


**FIGURE 8.67** Amplitude of $\lambda_2$ and frequency response curves with damping coefficients $c = 0{,}0.01, 0.015, 0.02$ for different fiber angle orientation (a) $\alpha = 0°$, (b) $\alpha = 30°$, (c) $\alpha = 60°$, (d) $\alpha = 90°$.

In Figures 8.68 and 8.69 the time-dependent behavior of $\lambda_{1,}$ $\lambda_2$ for different fiber angle orientation are displayed with the following assumptions $V_{ac}/V_{dc} = 0.1, 0.5$, $\tilde{V}_{dc}^2 = \sqrt{0.1}$ and $V_{ac}/V_{dc} = 0.1, 0.5$. In both Figures 8.68 and 8.69, the harmonic exciting frequency of ADE is considered $\Omega = 1.57$. By increasing the fiber angle orientation, it is evident that the max values $\lambda_1$ are so much higher than $\lambda_2$. Considering the damping coefficient $c = 0.01$, the maximal amplitude becomes smaller, and peaks of higher order are almost damped away. This happens since the damping force removes energy from the system, usually in the form of thermal energy and lead systematically to the stabilization of the system. Moreover, based on the results, a strong nonlinearity is also observed due to the oscillation of coupled $\lambda_{1,}$ $\lambda_2$. Results show that for $\alpha = 0°$ (Figure 8.69(a)) the amplitude $\lambda_2$ shows relatively less nonlinearly and exhibits a beating behavior. In contrast for $\lambda_1$ (Figure 8.68(a)) the opposite results are obtained and more nonlinearly behavior is depicted. According to the coupled nonlinear equations for $\lambda_{1,}$ $\lambda_2$, the results are not identical, and for various fiber directions, different behavior is concluded. The regular beating behavior is more seen in $\alpha = 30°$, $60°$ due to the inevitable effect of fibers in DE. As it was mentioned before, it can be concluded that as the damping has a small value but its effect in all directions is quite noticeable.

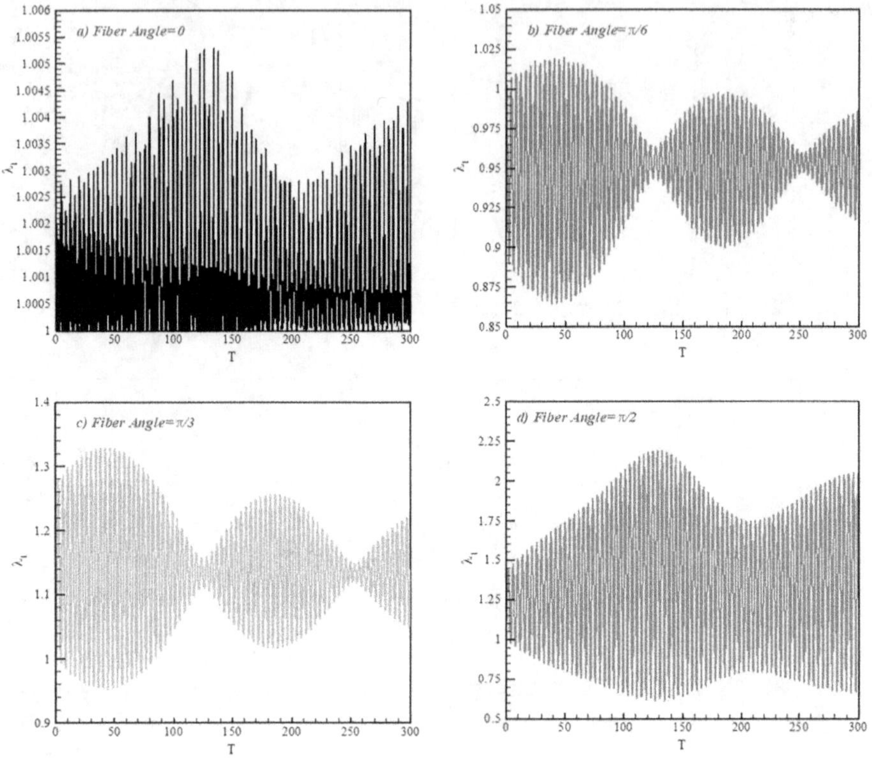

**FIGURE 8.68** Time-Dependent behaviors of $\lambda_1$ with $V_{ac}/V_{dc} = 0.1$.

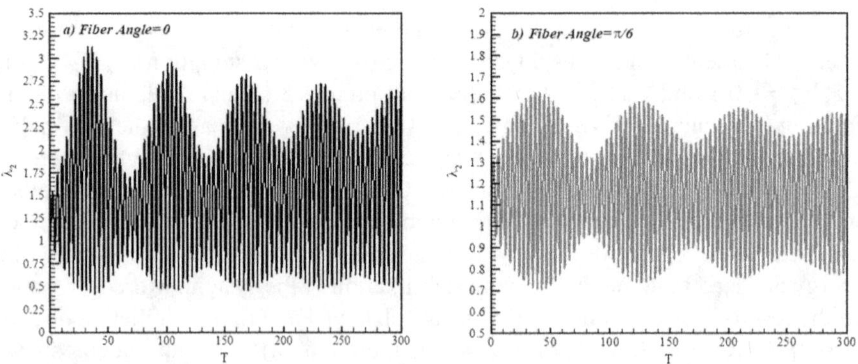


**FIGURE 8.69** Time-Dependent behaviors of $\lambda_2$ with $V_{ac}/V_{dc} = 0.5$.

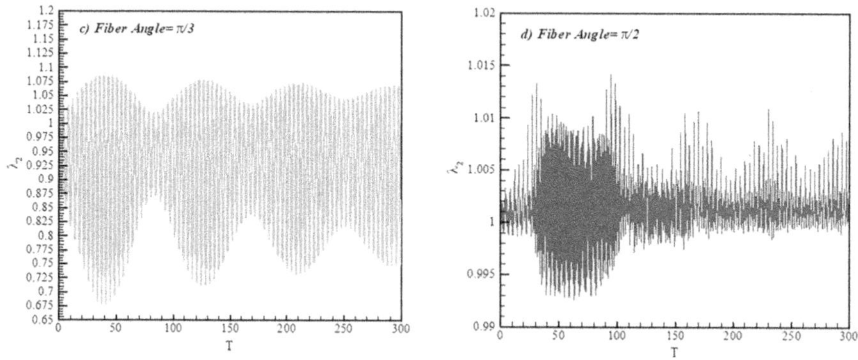


**FIGURE 8.69 (CONTINUED)** Time-Dependent behaviors of $\lambda_2$ with $V_{ac}/V_{dc} = 0.5$.

The phase diagrams of an ADE for different fibers orientation under the condition of with $V_{ac}/V_{dc} = 0.1, 0.5$, $\tilde{V}_{dc}^2 = \sqrt{0.1}$, $V_{ac}/V_{dc} = 0.1$, $\Omega = 1.35$, $c = 0.01$ and $\sigma_1 = \sigma_2 = 0.5$ are plotted in Figures 8.70 and 8.71. According to the curves of Figures 8.70 and 8.71 it is concluded that the ADE would experience highly complicated

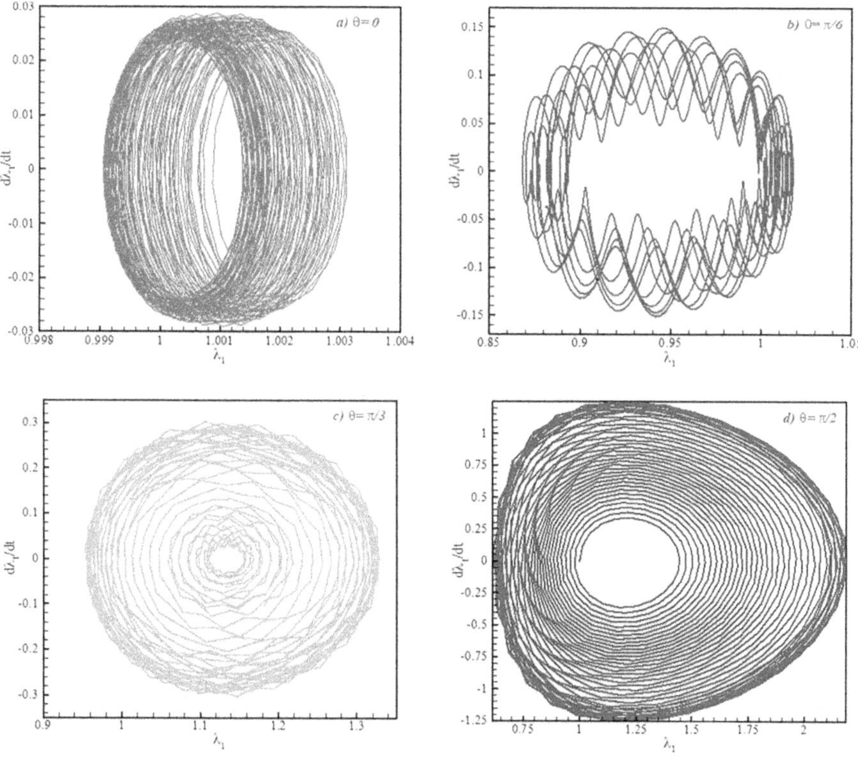


**FIGURE 8.70** Phase diagrams ($d\lambda_1 / dt - \lambda_1$) for different fibers orientation and $c = 0.01$.

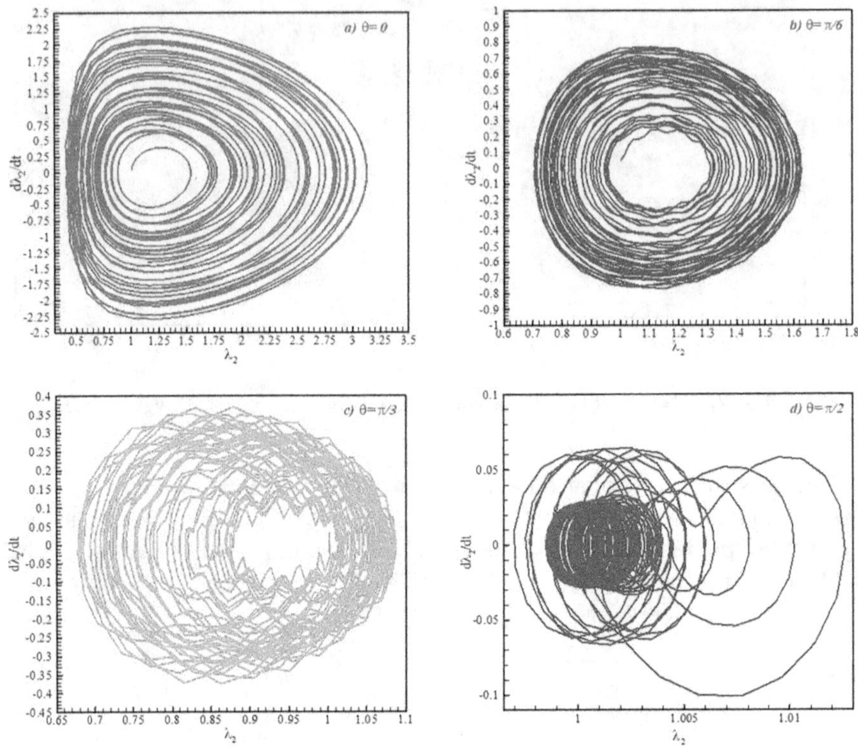


**FIGURE 8.71** Phase diagrams ($d\lambda_2 / dt - \lambda_2$) for different fibers orientation and $c = 0.01$.

nonlinear behavior. The curves are in almost closed regions and loops that show a stable situation for the dynamic oscillation of the ADE. It should be noted that considering the effect of damping causes irregularities in the behavior of the system. When $\alpha = 90°$ in Figure 8.71(d), disordered curves illustrate that the ADE undergoes an aperiodic vibration. This is where the effect of the fiber angle orientation comes into play. More results on the dynamic analysis of DEs are provided in [15] and [16].

## 8.6 CONCLUSION

This chapter finally presented the numerical results obtained from the proposed constitutive models. The comprehensive model was used in various forms to analyze various actuators, including single and multi-layered as planar and tubular actuators. First, we used the model including rate-dependency for a bending actuator consists of an isotropic inactive layer, attached to an anisotropic active one. Then the results are presented for anisotropic single layer tubular actuators under different loading conditions. Interesting findings inspired an innovative tubular actuator. Chapter 8 also covered the numerical results for some multi-layered fiber-reinforced actuators. This chapter also presented the calibration and validation procedures of the model, wherever it was required. Finally, because of importance of the dynamic quick response of smart materials with a dispersion-type anisotropy to the external electric field in smart manufacturing technologies, the nonlinear Lagrange equations were analyzed to present the useful and applicable results in this field. For different damping coefficients and fibers direction, a detailed study for the time-dependent dynamic deformation revealed the effect of fibers angle and damping coefficient on the dynamic behavior. Moreover, obtained phase diagrams depicted the stability transition and the nonlinearity of the system.

## REFERENCES

[1] L. D. Peel, *Fabrication and mechanics of fiber-reinforced elastomers*, Brigham Young University, 1998.

[2] J. Mehta, Y. Chandra, R. Tewari, The use of dielectric elastomer actuators for prosthetic, orthotic and bio-robotic applications, *Procedia Computer Science*, 133 (2018) 569–575.

[3] S. Son, N. Goulbourne, Dynamic response of tubular dielectric elastomer transducers, *International Journal of Solids and Structures*, 47(20) (2010) 2672–2679.

[4] A. Ahmadi, M. Asgari, Nonlinear coupled electro-mechanical behavior of a novel anisotropic fiber-reinforced dielectric elastomer, *International Journal of Non-Linear Mechanics*, 119 (2020) 103364.

[5] A. Ahmadi, M. Asgari, Novel bio-inspired variable stiffness soft actuator via fiber-reinforced dielectric elastomer, inspired by Octopus bimaculoides, *Intelligent Service Robotics*, 14(5) (2021) 691–705.

[6] E. Allahyari, M. Asgari, Fiber reinforcement characteristics of anisotropic dielectric elastomers: A constitutive modeling development, *Mechanics of Advanced Materials and Structures*, 29 (2021) 1–15.

[7] Z.M. Ghahfarokhi, M. Salmani-Tehrani, M.M. Zand, S. Esmaeilian, A new viscous potential function for developing the viscohyperelastic constitutive model for bovine liver tissue: Continuum formulation and finite element implementation, *International Journal of Applied Mechanics*, 12(03) (2020) 2050029.

[8] Y. Wang, B. Chen, Y. Bai, H. Wang, J. Zhou, Actuating dielectric elastomers in pure shear deformation by elastomeric conductors, *Applied Physics Letters*, 104(6) (2014) 064101.

[9] T. Vu-Cong, C. Jean-Mistral, A. Sylvestre, Impact of the nature of the compliant electrodes on the dielectric constant of acrylic and silicone electroactive polymers, *Smart Materials and Structures*, 21(10) (2012) 105036.

[10] O.A. Araromi, S.C. Burgess, A finite element approach for modelling multilayer unimorph dielectric elastomer actuators with inhomogeneous layer geometry, *Smart Materials and Structures*, 21(3) (2012) 032001.

[11] M. Majidi, M. Asgari, Nonlinear bending–twisting coupling in electromechanical finite deformation of fiber-reinforced tubular dielectric elastomer for soft actuators, *International Journal of Non-Linear Mechanics*, 156 (2023) 104480.

[12] F.S.C. Mustata, A. Mustata, Dielectric behaviour of some woven fabrics on the basis of natural cellulosic fibers, *Advances in Materials Science and Engineering*, 2014 (2014) 216548.

[13] M. Majidi, M. Asgari, Rate-dependent electromechanical behavior of anisotropic fiber-reinforced dielectric elastomer based on a nonlinear continuum approach: Modeling and implementation, *The European Physical Journal Plus*, 138(1) (2023) 1–29.
[14] M. Majidi, M. Asgari, Nonlinear electromechanical responses in multi-layered fiber-reinforced dielectric elastomer composites, *Thin-Walled Structures*, 197 (2024) 111599.
[15] E. Allahyari, M. Asgari, Effect of fibers configuration on nonlinear vibration of anisotropic dielectric elastomer membrane, *International Journal of Applied Mechanics*, 12(10) (2020) 2050114.
[16] E. Allahyari, M. Asgari, Nonlinear dynamic analysis of anisotropic fiber-reinforced dielectric elastomers: A mathematical approach, *Journal of Intelligent Material Systems and Structures*, 32(18–19) (2021) 2300–2324.

# Index

Pages in *italics* refer to figures and pages in **bold** refer to tables.

## S



## T



## U



## V



## W



## Y

For Product Safety Concerns and Information please contact our EU representative GPSR@taylorandfrancis.com
Taylor & Francis Verlag GmbH, Kaufingerstraße 24, 80331 München, Germany

www.ingramcontent.com/pod-product-compliance
Lightning Source LLC
LaVergne TN
LVHW020714110826
845149LV00012B/2268

* 9 7 8 1 0 3 2 9 4 5 6 9 9 *